TI-83+, TI84+ and TI-83+ Silver Graphing Calculator: How To Best Use It (3rd edition)
is the property of:

Adventures In Education, Inc.
3460 S Fletcher Ave Ste 205
Fernandina Beach, FL 32034
630-877-4006
adventuresinedu@earthlink.net
http://www.Adventures-In-Education.com

Editor: Kathleen L. Knowlton, Consultant, Glen Ellyn, IL
Proof Reader: Dr. Gail Pieper, Benedictine University, Lisle, IL

Trademark Acknowledgements:
TI® is a registered trademark of Texas Instruments Incorporated.

Copyright © 2015, 2011, 2005 by Adventures In Education, Inc.
All Rights Reserved. No part of this book may be reproduced or transmitted in any form or by any means, electronic or mechanical, including photocopying, recording, or by any information or retrieval system, without permission in writing from the published. The published hereby grants to individual teachers who have purchased this book permission to reproduce the activities and projects needed for use with their own students. Reproduction of the book for an entire school or school district or for commercial use is prohibited.

Printed in the United States of America.

To my husband Tom -

Thank you for your never-ending support, help and encouragement.

To All Educators:

The use of technology in the classroom is a very powerful tool. In many instances, calculator exercises prove to greatly enhance your instruction and allow you to explore more interesting and challenging problems. However, technology should not be used in lieu of teaching basic mathematical skills. Therefore, you will need to use your own judgment when to implement, substitute or apply this tool. The instructions, lessons and activities provided in this text are based on the use of the TI-83 Plus, TI-84 Plus, and TI-83 Plus Silver Edition Graphing Calculator. Any of these activities can easily be modified and adapted to the TI-83 Graphing Calculator. The TI-84 family is 100% compatible, keystroke for keystroke with the TI-83 Plus. I hope this text sparks many great ideas for your classroom. ENJOY!

Elisabeth Knowlton, PhD

FOREWORD

Elisabeth Knowlton received her Master's degree in Statistics from Northern Illinois University in 1993, PhD in Education from Northcentral University in 2011, and is the owner and founder of Adventures In Education, Inc. She has 20+ years of experience teaching mathematics at the college-level and is currently a faculty member at Portland State University and Benedictine University. Additionally, she provides educational and statistical consultation to publishing companies, various school districts and Midwestern University. Elisabeth's association with leaders in the educational and consulting field, as well as continuing classroom experience, places her at the forefront of new developments in education. Although Elisabeth performs at the university and college level, she routinely demonstrates the unique ability to reach children at all levels. She is equally adept at helping children ages 5 to 18 as she is at teaching adults enrolled in graduate level courses.

It has been my great pleasure to participate with Elisabeth in the development of *TI-83+, TI-84+ and TI-83+ Silver Graphing Calculator: How To Best Use It (3rd edition)*. It is gratifying to write this foreword acknowledging Elisabeth's outstanding achievements and contributions to the field of mathematics education. Elisabeth has also authored the following texts:

Knowlton, E. (2015). An Intervention Strategy Designed to Reduce Math Remediation Rates at Community Colleges. Fernandina Beach, FL: Adventures In Education, Inc.

Knowlton, E. (2014). Microsoft EXCEL® 2013: Essentials for Statistics and Data Management. Fernandina Beach, FL: Adventures In Education, Inc.

Knowlton, E. (2012). Quantitative Literacy: What Is It? How You Use It Each Day! Fernandina Beach, FL: Adventures In Education, Inc.

CONTENTS

UNIT A : Settings
- Mode 2-3
- Format 4-5
- Window 6-7
- Table 8-10
- Practice your skills 11

UNIT B: Miscellaneous
- Absolute Value 14
- Angles 15-17
- Order of Operation 17
- Some Tips 17
- Use your Catalog 18
- Diagnostics 19
- Exponents 20
- Fractions and Decimals 21
- Memory Storage and Retrieval 22
- Scientific Notation 23
- Engineering Notation 23
- Test 24
- Practice your skills 25

UNIT C : Graph - Draw – Zoom – Calc – Shade
- Graph 28
- Draw 29-34
- Zoom 35-39
- Calc 39-44
- Shade 45-46
- Practice your skills 47

**UNIT D: Lists – StatPlot – Stat Analysis
     Curve Fitting– Prob - Sim – Dist**
- Lists 50-52
- StatPlot 52-56
- Stat Analysis 57-59
- Curve Fitting 60-65
- Probability Computations 65-68
- Simulations 69-72
- Probability Distributions 73-81
- Practice Your Skills 82-83

UNIT E: Matrix – Matrix Op - Simult Equations – Other

Matrices	86-90
Matrix Operations	91-94
Solving Simultaneous Equations	95-100
More Matrix Operations	101-103
Practice Your Skills	104

UNIT F: Other Topics

Equation Solver	106
Polar and Parametric Functions	107-110
Store Data and Pictures	111-113
Transmit Data and Programs	114
Create a Sequence	115
Practice Your Skills	116

INDEX   117

UNIT A: Settings

Mode	**2-3**
Format	**4-5**
Window	**6-7**
Table	**8-10**
Practice your skills	**11**

A.1.: Mode

Check the settings of your **MODE** before you start. The following are the default settings of your calculator.

A.1.1.:Float: The number will be rounded to the number of decimal places chosen.

A.1.2.: Radian vs. **Degree** Setting: The default setting is in **Radian**. If you input numbers in radians, you don't have to worry.
Radian

Change to **Degree** if you input information in degrees.
Degree

Note: Make sure you change the setting back to the original once you are finished.

A.1.3.: Full Screen vs. Split Screen: Full Horiz G-T

Full: This means your calculator will graph with a full screen.
Enter your function in **Y=**, then press **GRAPH**

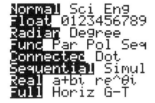  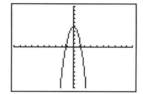

Horiz: If you set your mode to **Horiz**, your calculator will split the screen horizontally as shown below: Use the same function as above, but watch what will happen! How did you input your function? Did you press **Y=** or **GRAPH**?

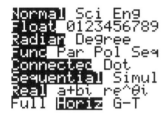 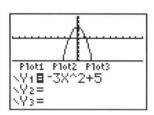

Maximum value for a row is 30.
Maximum value for a column is 94.

G-T: If you set your mode to **G-T**, the screen will be split vertically as follows:

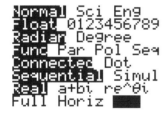  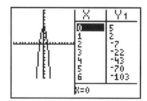

You probably noticed that instead of the **Y=** screen, the table setting appeared. We will investigate tables and their properties later.

Note: We will also return to the other items on the **Mode** menu later in this course.

A.2.: FORMAT

To display the format settings, press **FORMAT** or **2ⁿᵈ ZOOM**. The default settings for **FORMAT** are as follows:

A.2.1.: RectGC PolarGC   Sets cursor coordinates

RectGC - rectangular graphing coordinates
   RectGC and **CoordOn** - X and Y are displayed

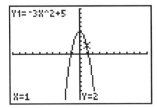

PolarGC – polar graphing coordinates
   PolarGC and **CoordOn** – R and θ are displayed

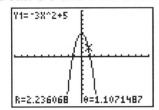

A.2.2.: GridOff GridOn
   GridOff - does not display grid points   **GridOn** – displays grid points

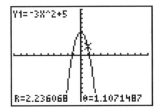   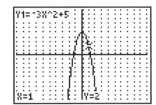

A.2.3.: AxesOn AxesOff

AxesOn – displays the axes

FORMAT   GRAPH   TRACE

AxesOff – does not display the axes.

FORMAT   GRAPH   TRACE

This setting is usually chosen if you would like to create a picture on your screen using a program. For example: HAPPY BIRTHDAY TO YOU. I will share the program with you later!

This program was created with the following **FORMAT** settings: **RectGC, CoordOff, GridOff, AxesOff, LabelOff, ExprOff**.

A.2.4.: LabelOn LabelOff - displays labels for X and Y axes (or likewise r and θ)
ExprOn ExprOff - determines whether to display the Y=expression when the trace cursor is activated.

FORMAT   GRAPH   TRACE

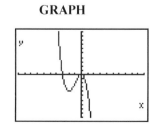
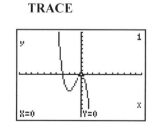

A.3.: WINDOW

To display the window settings, press **WINDOW**.

Window with **Func** setting:
The default **WINDOW** with the default **MODE** settings are as follows:

Window with **Par** setting:
If you change the **Mode** setting to **Par** you will get the following default settings:

Window with **Pol** setting
If you change the **Mode** setting to **Pol** you will get the following default settings:

 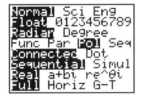

Window with **Seq** setting
If you change the **Mode** setting to **Seq** you will get the following default settings:

 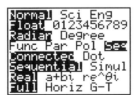

Note: We will apply and practice the application of different settings in exercises throughout the course.

Change WINDOW setting:

Before graphing a function, check your **WINDOW** settings. The **WINDOW** determines the dimensions and scale of you coordinate system. For a **standard (default) WINDOW** set the length of the X-Axis and Y-Axis using the cursor:

<u>For example</u>: Change the **WINDOW** settings to X =[-20,20], X-scale = 5, Y=[-6,6], Y-scale=2. *Note*: use the **(-)** key to input negative numbers. Your graph will now be displayed with the following units:

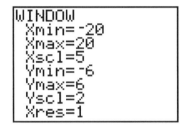 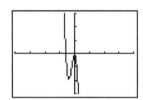 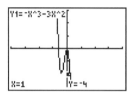

Overview for all WINDOW settings:

A display of all **WINDOW** settings for different modes are shown below:

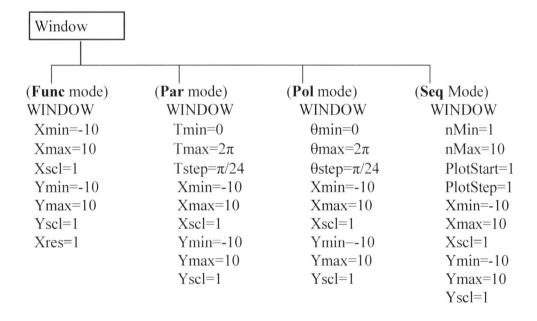

A.4.: TABLE

A.4.1.: To display the table settings, press **TBLSET** or **2nd WINDOW**.

This screen lets you set the starting value of the table **TblSet =** as well as the increments **ΔTbl=** .

Example: If you would like to get the values for the function y = 2x -5 starting at -3 at increments of .5 units, then input the following in **TblSet** and **Y=** screens:

Then press **TABLE** or **2nd GRAPH**:

Use your arrow key to scroll down. What do you notice?

Auto vs. Ask:
You may wonder what these options mean. The Auto option will automatically produce a table with the above specifications. The Ask option will let you input a value for the Independent variable and not create a table with all the values automatically. Let's try this.

A.4.2.: Find Values for a Function:

Use the following Table Setup **TBLSET** or **2nd WINDOW,** the same function as above and press **TABLE** or **2nd GRAPH**:

    TBLSET   **Y=**   **TABLE**

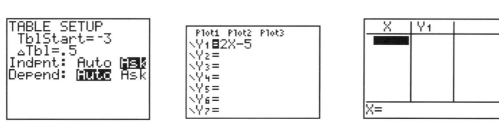

You will get an empty table. Now input the desired values for X to compute Y1. The Y1 values will be appearing.

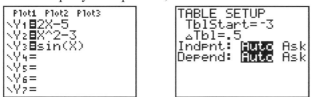

A.4.3.: Find Values for more than one Function:

You can compute values for more than one function. For example you can have numerous functions such as Y1, Y2, Y3 etc. and the table will compute the values for all the functions. For example you input Y1, Y2 and Y3 in **Y=** :

If you would like to compute the y-values for all three functions automatically and simultaneously, just press **TABLE** or **2nd GRAPH** and you will see the following:

But where is Y3? Scroll to the right!

How many Y's can you input? You can compute y-values for Y1 through Y9, and Y0.

Number of Dependent Variables possible in TABLE display for various graphing modes:

Graphing Mode	Independent Variable	Dependent Variable
Func (function)	X	Y1 through Y9, and Y0
Par (parametric)	T	X_{1T}/Y_{1T} through X_{6T}/Y_{6T}
Pol (polar)	θ	r_1 through r_6
Seq (sequence)	n	u(n), v(n), and w(n)

A.4.4.: G-T Split: Graph-Table Split Screen

You have seen the split screen option in the previous chapter. Remember the G-T screen which displayed the graph and the table. Let's see what will happen if we change our **MODE** to **G-T** and press **TABLE**. Assume the same functions and table settings as above.

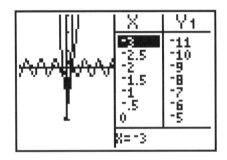

Does it matter if we press **GRAPH** or **TABLE** to display the **G-T** screen?

Where would you find the values for Y2 and Y3?
Press TABLE and use your cursor to scroll to the right. You will get the following:

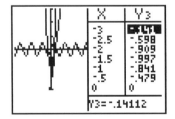

Practice Your Skills UNIT A

1. Given the function Y1 = -2x + 5 and Y2 = x^2 -2x -8.
   a. Find the value of Y1 for x = -1, 0, and 5.
   b. Find where Y1 and Y2 are 0.

2. The given polynomial function can be used to estimate the number of milligrams of ibuprofen in the bloodstream at t hours after 400 mg of medication has been swallowed.
   M(t) = 0.5 t^4 + 3.45 t^3 – 98.65 t^2 + 347.7 t, 0 < t < 6

   a. Graph the function using the viewing window [-24, 15] Xscl = 3, [-14,400, 8000] Yscl = 800. Sketch the graph.
   b. Graph the function using the viewing window [0, 6] Xscl = 1, [-200, 500] Yscl = 100. Sketch the graph.
   c. Which graph would you use to predict the blood content of ibuprofen after t=5 hours?

3. Draw a Cat:
   Input: Y1 = √(.5x – 6) +4
   Y2 = √(-.5x – 6) +4
   Y3 = - √(.5x – 6) +4
   Y4 = - √(-.5x – 6) +4
   Y5 = (- abs(-x) + 3) (x ≥-3) (x ≤ 3)
   Y6 = - √(4 - .2x^2) – 3

   a. Set your Mode and Format so that you can only see the cat. Sketch the graph:

   b. Which window did you use?

   c. Use your x and y scale to determine the domain of the above functions.

   d. Use your table function to find the domain of the above functions.
      Domain for
      Y1 =
      Y2 =
      Y3 =
      Y4 =
      Y5 =

UNIT B: Miscellaneous

Absolute Value	14
Angles	15-17
Order of Operation	17
Some Tips	17
Use your Catalog	18
Diagnostics	19
Exponents	20
Fractions and Decimals	21
Memory Storage and Retrieval	22
Scientific Notation	23
Engineering Notation	23
Test	24
Practice your skills	25

B.1.: Absolute Value

Find |-4.236| =

Method 1: Use MATH menu
Select **MATH** and **NUM**, then select **1:abs(** and press **ENTER**, input number **-4.236)** (*Note*: you have to use **(-)**) and press **ENTER**.

 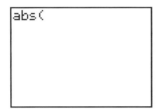 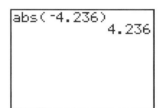

Method 2: Use CATALOG menu
Press **2nd** and **0**, the arrow points to **abs(** . Press **ENTER** and input number **-4.236)** and press **ENTER**. (*Note*: you have to use **(-)**).

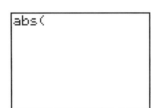

Practice:
1. Find |-7+5| =

2. Find - | -5| =

3. Graph y = |x – 3|

B.2.: Angles

Setting on your calculator: **ANGLE**
    1: °   degree symbol
    2: '   degree minute
    3. r   radian symbol
    4: >DMS   change to degree-minute-second form
    5: R>Pr(change to polar radian measure
    6: R>Pθ(
    7: P>Rx(
    8: P>Ry(

Set your **MODE** to **Radian** and press **2nd MODE** (or **QUIT**) to return to home screen.

Enter angles in degrees and minutes:
To input 85°45': type **85 ... ANGLE** (or **2nd APPS**) ... **1: °** ... **ENTER,** then type
45 ... ANGLE (or **2nd APPS**) ... **2: '** .... **ENTER ENTER**. It will convert your angle from degrees and minutes to an angle in degrees with a decimal.

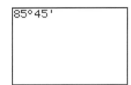 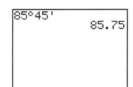

Change angle in degrees and minutes **to angle in degrees, minutes, seconds** enter:
type **85... ANGLE** (or **2nd APPS**) ... **1: °** ... **ENTER,** then type
45.5 ... ANGLE (or **2nd APPS**) ... **2: '** ... **ENTER,**
press **ANGLE** (or **2nd APPS**) ... **4:>DMS** ... **ENTER**

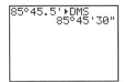

DEGREES to RADIANS

Set your **MODE** to **Radian** and press **2ⁿᵈ MODE** (or **QUIT**):
To change 25° to radians: type **25 … ANGLE (or 2ⁿᵈ APPS) … 1:° … ENTER**.

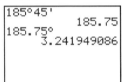

Note: If you enter an angle in degrees and minutes such as 185°45':
Type **185 … ANGLE (or 2ⁿᵈ APPS) … 1:° … ENTER**.
To change to radians type **45 … ANGLE (or 2ⁿᵈ APPS) … 2:' … ENTER ENTER. What happened? Did the conversion take place? What do you need to input?**

RADIANS to DEGREES

Set your **MODE** to **Degree** and press **2ⁿᵈ MODE** (or **QUIT**):
Watch what happens!!

The calculator assumes you input the angle in degrees because you set your **MODE** to **Degree**.
If you want to change an angle in radian measure to degree you need to do the following:
Change 2π to degrees: Type **2π … ANGLE (or 2ⁿᵈ APPS) … 3:ʳ … ENTER … ENTER**.
Let's see what 1 radian will equal: Type **1 … ANGLE (or 2ⁿᵈ APPS) … 3:ʳ … ENTER … ENTER**.
What will π/2 equal?

Note: You probably realized by now that you have to be very careful working with angles. My advice is to just leave your settings on **Radian**. If you work with degrees, change the angle to radians first and use that value in your computations.

B.3.: Order of Operation

In the following expression, if there is more than one operation in the numerator or denominator, they must both be enclosed in parentheses to ensure correct order of operations. Your calculator does not perform correct order of operation automatically.

$$\frac{10 \div (5-2) \cdot 2 + 9 \cdot 4}{3^5 + 2^3}$$

Type **(10 ÷ (5 – 2) x 2 + 9 x 4) ÷ (3^5 + 2^3) ENTER** = .1699867168

Note: The TI-83 can only use parentheses in an order of operations, it will not recognize brackets [ ] or braces { } as symbols of grouping.

$5\{3 + 2[5 – 6(3 + 2) – 7] – 8\}$ must be entered as …

5(3 + 2 (5 – 6(3 + 2) – 7) – 8) = -345

Try: $\frac{18 \div 5 + 3\,(3)}{22.35 + 99.12 - (3^2 + 7^2)}$

Answer: .1985189853

Some tips:

- If you make an error or omit a character before you press **ENTER**, you may arrow back to the error and press **DEL** to delete the extra character or press **2nd INS** to insert a forgotten character or just go to the error and overwrite the incorrect character.
- If you made an error after pressing ENTER, press 2nd ENTRY. This reiterates past entries … how many depends on the number of characters previously used.
- Be sure to use the (-) for negative and - for subtraction: -2 - 5 = **(-) 2 - 5**

B.4.: Using your Catalog

The Catalog is an alphabetic list of all functions and instructions on the TI-83 Plus. Some of these items are also available on keys and menus.

To select from the Catalog:
1.) Press **CATALOG** (or **2nd 0**).
2.) Use **down arrow and up arrow** to move the cursor to the desired function or instruction.
3.) Press **ENTER** and your selection will be shown on the home screen.

Example: You would like to find **ref**.
1. **CATALOG**
2. **ALPHA R**
3. use down arrow key to select **ref**.
4. **ENTER**

B.5.: Diagnostics

The displayed values for r (the correlation coefficient) and/or r^2 (the coefficient of determination) may be switched on or off by using the **DiagnosticOn** or the **DiagnosticOff** instruction.

DiagnosticOff
If the values for r and/or r^2 do not need to be displayed, from the home screen press 2nd ... 0 ... x^{-1} and scroll down to the option **DiagnosticOff.** Press **ENTER ENTER**. The word **Done** should appear on the home screen.

The displayed values for a linear regression model will not show r and r^2

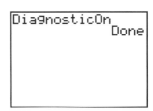

DiagnosticOn
If the values for r and/or r^2 need to be displayed, from the home screen press 2nd ... 0 ... x^{-1} and scroll down to the option **DiagnosticOn.** Press **ENTER ENTER**. The word **Done** should appear on the home screen. Linear Regression output will now show r and r^2.

B.6.: Exponents

To take a number to the 2nd power such as 4^2 type
   a. **4^2** = 16
   b. **4 ... x²** = 16

Use MATH menu:

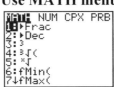

3rd Power: To take a number to the 3rd power such as 4^3 type: **4 ... MATH ...3: ³ ... ENTER** = 64

4th Power and any other power: To take a number to the 4th power such as 2^4 type: **2^4 ... ENTER** = 16

Fractional Exponent: To enter a fractional exponent, always use parenthesis such as $25^{1/2}$ type: **25 ^ (1/2) ... ENTER** = 5

3rd Root: To find the 3rd root of a number such as $27^{1/3}$ or $\sqrt[3]{27}$ use:
MATH ... 4: ³√ (... 27) ENTER = 3

Nth Root: To find the nth root of a number such as $32^{1/5}$ or $\sqrt[5]{32}$ use:
Type **5 ... MATH ... 5: ⁿ√ (... 32) ENTER** = 2

B.7.: Fractions and Decimals

Use MATH menu:

Add Fractions:
1/2 + 1/3 =
Type **1/2 + 1/3 ... MATH ... 1:>Frac ... ENTER** = 5/6

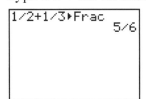

Express a decimal as a fraction:
.5 =
.42 =
.333333 =
Type **.5 ... MATH ... 1:>Frac ... ENTER** = 1/2, etc.

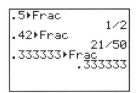

You can use **>Frac** with real or complex numbers, expressions, lists and matrices.

Note: If the answer cannot be simplified or the resulting denominator is more than three digits, the decimal equivalent is returned.

Express a fraction as a decimal:
1/2 =
21/50 =
1/3 =
Type **1/2 ... MATH ... 2:>Dec ... ENTER** = .5 etc

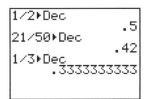

You can use **>Dec** with real or complex numbers, expressions, lists and matrices.

B.8.: Memory storage and retrieval

To store a value in the memory of the TI-83, the value must be given a variable name

For example: To store the value 6.123 into the memory:
Type **6.123 ... STO> ... ALPHA A** (or the letter of your choice) **...ENTER**
To recall the value of **A** press **ALPHA A** again.

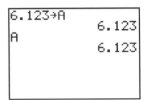

B.9.: Check Available Memory

Use **MEMORY** menu: **MEM** (or **2nd +**)

TI83 Silver

2: Mem Mgmt/Del: Check RAM and ARC (Archive) memory availability/usage

TI83 Plus

1: Check RAM: Reports memory availability/usage

B.10.: Scientific Notation

Input 375×10^{15} in scientific notation:

Method 1: Use **EE** Mode
To enter a number in scientific notation press 2nd ... ' to access EE. A number is entered as 375 2nd ... ' then enter **15**. The number will be converted to 3.75 E 17 .

Method 2: Change **MODE** to **Sci**

Input 375×10^{15} in scientific notation. Change MODE to Sci, then press 2nd **MODE** (or **QUIT**) to get back to your home screen. Type **375 * 10^15** and press **ENTER**. What do you see?

 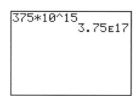

Input 75,000,000 in scientific notation. Change MODE to Sci, then press 2nd **MODE** or **QUIT** to get back to your home screen. Type **75,000,000** and press **ENTER**. TADAAA!!

 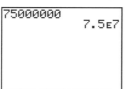

Engineering Notation

Change **MODE** to **Eng**. Eng notation will express any number in powers of 3, 6, 9, etc. Then press 2nd **MODE** or **QUIT** to get back to the home screen. Input: 750 and you will get 750E0, 75,000 and you will get 75E3, 240,000,000 and you will get 240E6. Try 10,100,000,000 ...

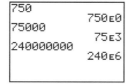

B.11.: Test

To test a mathematical statement, compare lists or apply logic, use the **TEST** menu. You can access this menu by pressing **TEST (or 2nd MATH)**. The following screen will appear:

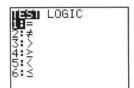 using your right arrow: **LOGIC** will appear

Note: LOGIC is used mostly for programming.

TEST

If the relationship is true, then a **1** will appear. If the statement is false, a **0** will appear.

Compare Numbers and Algebraic Expressions:

Let's see how this works:

Test $3 < -2$. We know that it is false, but let the calculator confirm this.

Type **3** and select **TEST** or **2nd MATH**, select **5: <** and press **ENTER**,
input **-2** (use the (-) sign!!) and press **ENTER**.

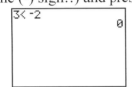

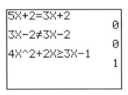

Test other relationships. Can you explain the results?

Compare Lists:

Again if the relationship is true, a **1** will appear. If the statement is false, a **0** will appear.
It will compare each element of the list one by one.

Let's test the following lists: $\{2,4,6\} < \{3,4,5\}$

Input the list as follows:

Input **{** or **2nd (** , enter **2, 4, 6 }**
TEST or **2nd MATH** and choose **5:<**, press **ENTER**
Input **{** or **2nd (** , enter **3, 4, 5 }** and press **ENTER**

You will see: {1 0 0} which means the relationship of the first items of the lists is true but the others are false:

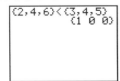

Practice your skills UNIT B
(Miscellaneous Topics)

1. $-12/6 =$
2. Ans $+ 11 =$
3. 13^2 (Use the x^2 key) =
4. $13\wedge 2 =$
5. $\sqrt{40} =$
6. $5 * \sqrt[5]{32} =$
7. $11/33 =$ (express as a fraction)
8. $|-4-3| =$
9. $24/3*4 =$ (follow order of operations!!)
10. $(24/3)*4 =$
11. $24/(3*4) =$
12. $3 + 1/5 =$ (express as a fraction)
13. $1 - 1/6 =$ (express as a decimal)
14. $2/3 + 1/2 =$ (express as a fraction)
15. $3\ 1/5 + 1\ 1/6 =$ (express as a decimal)
16. Test: $5/8 < 2/3$
17. Test: $5/8 > 4/7$
18. Test: $3/4 = 7/9$

With calculator in Normal mode:
19. $2 * 10^4 =$
20. $2.5 * 10^4 =$
21. $2 * 10^{-4} =$
22. $2.5 * 10^{-4} =$
23. $(3.2 * 10^{-5}) * (4.8 * 10^7) =$
24. $35000000 * 6.8\ E\ 6 =$

With calculator in Sci mode:
25. $275000000 =$
26. $0.0000025 =$
27. $2 * 10^4 =$
28. $35000000 * 6800000 =$

Angles:
29. Convert to Degrees-Min-Seconds: $45.2° =$
30. Convert to Degrees: 1.8 radians =
31. Convert to Degrees: $(2/3)\pi =$
32. Convert to Radians: $145° =$
33. Cos $145° =$
34. Sin $1.34 =$
35. Sin $(\pi/2) =$

UNIT C : Graph - Draw – Zoom – Calc – Shade

Graph	**28**
Draw	**29-34**
Zoom	**35-39**
Calc	**39-44**
Shade	**45-46**
Practice your skills	**47**

C.1.: GRAPH and DRAW

C.1.1.: GRAPH
To graph a function its equation must be entered in **Y=** editor.

1. Type the equations: **Y1** = $x^2 - 4$
                                    Y2 = $3x - 2$
Press **GRAPH**.

 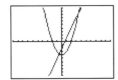

- Check your window: press **WINDOW**

Adjust WINDOW to be able to see both intersection points....

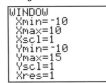 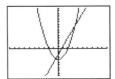

This will do!

General Guidelines: What to check before you graph?
- Check your **MODE**: set MODE to standard settings
- Check your **WINDOW**: use standard settings at first
- Check if **Stat Plots** are turned off: If **Plot1** is blinking then the stat plot is on. Move your cursor to **Plot1** and press **ENTER**. This will turn off your Plot1. Otherwise, it may interfere with your graph.

                       Plot1 on   Plot1 off

---

2. Graph: $y = 4x^3 - x^4$. Which window should you use to be able to see the maximum of the graph?

3. Graph: **Y1** = $e^{3.2} - 2x$

             Y2 = $\dfrac{3}{(1.01x - 3)^2} + 2$

Do these two graphs intersect? What window should you use?

C.1.2.: DRAW

To activate the **DRAW** menu press **DRAW (or 2nd PRGM)** and the menu below will appear. The graph will draw on the x-y coordinate system with the settings you selected.

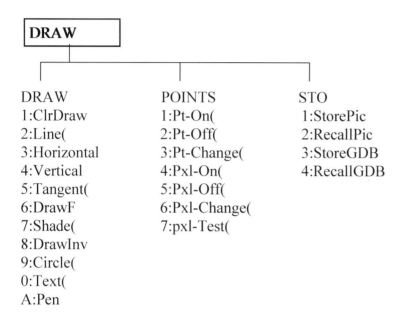

DRAW
1:ClrDraw
2:Line(
3:Horizontal
4:Vertical
5:Tangent(
6:DrawF
7:Shade(
8:DrawInv
9:Circle(
0:Text(
A:Pen

POINTS
1:Pt-On(
2:Pt-Off(
3:Pt-Change(
4:Pxl-On(
5:Pxl-Off(
6:Pxl-Change(
7:pxl-Test(

STO
1:StorePic
2:RecallPic
3:StoreGDB
4:RecallGDB

Note: Before you start clear existing drawings with **ClrDraw**.

C.1.2.1: ClrDraw
This function will clear all drawn elements. To clear drawings on a graph from the home screen or a program, start with a blank line on the home screen or in the program editor. Press **DRAW ... 1:ClrDraw... Enter.**

C.1.2.2: 2:Line(

Draw a Line:
Method 1: This function draws a line segment between 2 points and uses the coordinates (X1,Y1) and (X2,Y2) in the following order **Line(X1,Y1,X2,Y2)**.

Draw a line between the points (-2,1) = (X1, Y1) and (5,-3) = (X2, Y2).
Select **DRAW ... 2:Line ... Line(-2,1,5,-3) ... ENTER**. The graph will appear.

 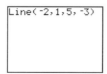 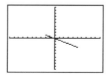

Method 2: Select two points directly on the coordinate system.
- Press **GRAPH.**
- Select **2:Line(** from the **DRAW** menu, press **ENTER**.
- Place the cursor on the desired point where you want your line segment to begin, press **ENTER**
- Move the cursor to the other point you want the line segment to end. The line is displayed as you move the cursor.

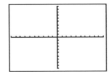  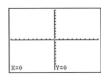 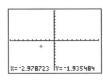 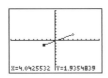

Erase a line segment:

Enter the parts you would like to erase: **Line(X1,Y1,X2,Y2,0)**
Example: You entered the line **Line(0,0,6,9)** and would like to erase parts in the middle **Line (2,3,4,6,0)**

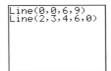 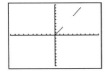

C.1.2.3.: Horizontal Line
Draw a horizontal line use **DRAW ... 3:Horizontal** ...input the value ... **ENTER**

 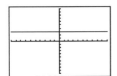

C.1.2.4.: Vertical Line
Draw a vertical line use **DRAW ... 4:Vertical** ...input the value ... **ENTER**

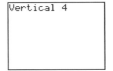 

C.1.2.5.: Tangent Line

1. To <u>draw a tangent line</u> to an existing graph use the **5:Tangent(expression, value)** function.
If you would like to draw a tangent line to **Y1= x^2** at x = 2, just enter **5:Tangent(Y1, 2)**. This will graph a tangent line to **Y1** at X=2.
(To enter **Y1** use **VARS** ... select **Y-VARS** ... **ENTER**... select **FUNCTION 1:Y1**... **ENTER**.)

 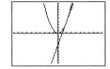

2. <u>Equation of a tangent line</u>: Suppose you want to find the equation of the tangent line at X=√2 / 2 for the function Y=cos x.
a. **Y1=cos(x) ... GRAPH** to Graph Y1=cos(x) in the equation editor and set **MODE** to 3 decimal places.
b. **ZOOM...7:Trig** to graph in Trig mode
c. **DRAW (or 2nd PRGM) ... 5:Tangent(** ... to initiate tangent instruction.
d. Enter √2 /2 for x.
e. Press **ENTER**.

    (d)   (e)

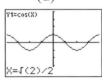

The tangent line is drawn at the X value and the equation of the tangent line is displayed on the screen: y = -.65x + 1.22

C.1.2.6.: Draw a Function
DrawF draws a function in terms of X on the current graph. This will produce the same graph as the graph editor **Y=** . This function **DrawF** also allows you to shift the graph up or down.

1. <u>If you want to graph</u> Y= $2x^2 - 1$ using the **DrawF** function.
    a. **DRAW ... 1:ClrDraw ... ENTER ...ENTER** to clear all drawings.
    b. Input Y1=$2x^2 - 1$ in **Y=**.
    c. **DRAW ... 6:DrawF...ENTER**
    d. Input **Y1**
        a. (To enter **Y1** use **VARS** ... select **Y-VARS** ... **ENTER**... select **FUNCTION 1:Y1**... **ENTER**.)
    e. **ENTER**

Your graph will appear.

 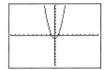

2. Graph: Y1 + 6 : (shift graph up 6 units)
If you would like to graph Y1 + 6, use the above steps but in step 4
   a. Input **Y1 + 6**
   b. **ENTER**

This will shift the graph 6 units up.

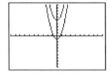

3. Draw $y = x^3 - 4$, shift the graph 3 units down and shift the graph 5 units up.
4. Draw $y = \sqrt{(12 - x^2)}$ and shift the graph 4 units down.

C.1.2.7.: Draw Inverse (works in **Func** mode only!)

Select **8:DrawInv(** from the **DRAW** menu.
Draws the inverse of a function or expression by plotting **X** values on the y-axis and **Y** values on the x-axis. However, you will not be able to get the function of the inverse.

1. To draw the inverse of $y = x^3 + 2$ input
   a. **Y1 = x³ +2** in the equation editor Y= .
   b. Press **GRAPH** to graph the function.
   c. **DRAW … 8:DrawInv …ENTER**.
   d. Type variable **Y1… ENTER**
   The graph of your Y1 and its inverse will show.

2. Draw the inverse of $y = -5x + 2$
3. Draw the inverse of $y = \cos x$
4. Draw the inverse of $y = \sqrt{x}$

C.1.2.8.: Draw a Circle (set your **ZOOM** to **5:ZSquare** to make a circle appear round)

Use **DRAW... 9:Circle(** ... to draw a circle with center (X,Y) and radius: **Circle(X,Y,radius)**

1. Draw a circle at (0,0) with radius 4.
   Use **DRAW... 9:Circle(0,0,4) ...ENTER**

 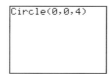 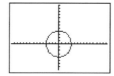

2. Draw a circle at (-2,4) with radius 5.3.
   Use **DRAW... 9:Circle(-2,4,5.3) ...ENTER**

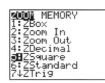 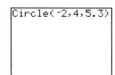 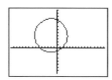

3. Draw a circle at (8.45, -6.78,20). Find an appropriate window so that the entire circle will appear round on the graph.

C.1.2.9.: Place Text on a graph screen

1. **Place text directly on a graph**:
   a. Create a graph first, i.e. **Y1 = $x^3 - 2x + 4$**
   b. **DRAW**
   c. select **0:Text(**
   d. Place cursor at the desired location on your graph where you would like to begin drawing and type '**Y1 = $x^3 - 2x + 4$**'

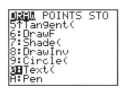 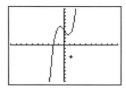 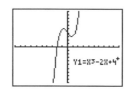

To cancel drawing: Press **DRAW** and **1:ClrDraw**

2. Place text on a graph from the home screen or a program:
   a. **DRAW**
   b. Select **0:Text(row,col,value)**

Rows are from 0 to 57 starting at the top left corner.
Columns range from 0 to 94.

Example: To place the text "Y1 = x^3 – 2x + 4" at row 5 and column 10.
1. **DRAW**
2. Select **0:Text(5,10,"Y1= x^3 – 2x + 4")**
3. **ENTER**
4. **GRAPH**

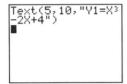 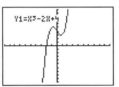

C.2.: ZOOM

Press **ZOOM** to view the zoom menu. Press ↓ to view more options.

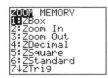 

MEMORY Menu

1: ZBox: defines a viewing window by selecting the corners of the box.
2: Zoom In: displays a smaller viewing area of the graph; thus magnifying the current graph..
3: Zoom Out: displays a larger viewing area of the graph; thus reducing the current graph..

4: ZDecimal: sets Xmin= -4.7, Xmax= 4.7, Ymin= -3.1 and Ymax= 3.1. Each pixel represents 0.1 units. The tic marks on each axis are equally spaced.

5: ZSquare: adjusts Xmin and Xmax so that the tic marks on the x axis are spaced the same as those on the y axis. A circle will appear round.

6: ZStandard: sets Xmin= -10, Xmax= 10, Xscl= 1, Ymin= -10, Ymax= 10 and Yscl=1.

7: ZTrig:
in radian mode: sets Xmin= -6.152285 about 2π, Xmax= 6.152285 about 2π, Xscl= 1.507963 about $\pi/2$, Ymin= -4 and Ymax= 4 and Yscl=1 and
in degree mode: sets Xmin= -352.5, Xmax= 352.5, Xscl= 90, Ymin= -4 and Ymax= 4 and Yscl=1

8: ZInteger: sets Xscl = 10 and Yscl = 10, and centers the graph at the cursor location.
9: ZoomStat: defines the viewing window so that all data points are shown (use for stat plot).
0: ZoomFit: adjusts Ymin and Ymax to include the minimum and maximum y values for the selected functions.

To return to the standard screen: Press **ZOOM** key and choose **6:ZStandard**
To return to the previous ZOOM screen: Press the **ZOOM** key and arrow right to **MEMORY,** select **1:ZPrevious** will take you back to each previous screen.

C.2.1.: 1: ZBox

It is possible to "zoom-in" on a portion of a graph by building a Z-box.

1. Draw the graph $Y1 = 4x^3 - 3x^4 + 4$. If you would like to look a little closer at the maximum point,
- activate the **ZOOM** menu select **1:ZBox** (c)
- the graph will reappear with a flashing cursor on the screen at the (0,0) location.
- Use the arrow keys to move the cursor to the location where you would like to build the box
- Press **ENTER**
- Arrow right or left then down or up to build the box (d)
- Press **ENTER** - the section of the graph that you've boxed in will appear (e)
- Select **TRACE** if you like to trace the graph (find coordinates) (f)
- To return to the original setting press **ZOOM** … select **6:ZStandard** … **ENTER**

a)

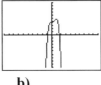
b)

c)

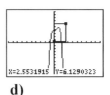
d)

e)

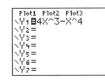
f)

2. Draw the graph $Y1 = 4x^3 - x^4$. The viewing screen does not show the entire graph.

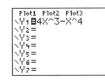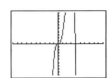

How can you view the entire graph and zoom in on the maximum point?

3. Use a Z-box to determine if the two graphs below intersect:
$Y1 = -x^2 - 4x - 6$ and $Y2 = x^2 + 2x - 1.2$

C.2.2.: 2:Zoom In

Using the **Zoom In** function will magnify the portion of the graph around the current cursor location. How much? The factor **MEMORY … 4:Set Factors** by which the graph is magnified is determined by the x factor and y factor settings.

- Graph the function **Y1= $-x^2 - 4x - 6$**
- Move your cursor to the desired point
- Press **ZOOM** … select **2:ZoomIn …ENTER … ENTER**
- Your graph is now magnified with the cursor in the center.

To return to the normal screen: Press the **ZOOM** key and choose **6:ZStandard**

 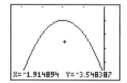

C.2.3.: 3:Zoom Out

Using the **Zoom Out** function will reduce the portion of the graph around the current cursor location. Graph the function Y1 = **Y1= $-x^2 - 4x - 6$**

- Move your cursor to the desired point
- Press **ZOOM** … select **3:ZoomOut …ENTER … ENTER**
- Your graph is now reduced with the cursor in the center.

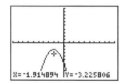 

C.2.4.: 4:ZDecimal

Using the **ZOOM … 4:ZDecimal** you will get the following graph

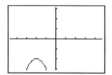

C.2.5.: 5:ZSquare

This will allow you to make a circle appear square.
Graph: $Y1 = \sqrt{25 - x^2}$ and $Y2 = -\sqrt{25 - x^2}$
This graphs the circle $x^2 + y^2 = 25$

6:ZStandard   **5:ZSquare**

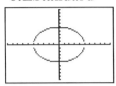 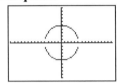

C.2.6.: 6: Zstandard

This will use the window X = (-10,10) scl=1, Y=(-10,10) scl=1

C.2.7.: 7:ZTrig

Graph $Y1 = 2 \sin x$

6:ZStandard   **7:ZTrig**

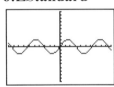 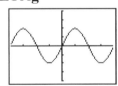

C.2.8.: SetFactors

To set Zoom Factors to determine the factor by which the Zoom In and Zoom Out options magnify or reduce a graph, press **ZOOM**, select **MEMORY** and **4:SetFactors**

The default factors are 4. This means that the window dimensions change by a factor of 4 when zooming in or zooming out. For a standard window zooming in will change the dimensions to:

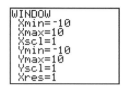 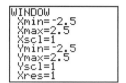

The zoom factors can be changed to any value greater than 1, but they do not have to be integers. These factors and the current cursor position determine the new window settings.

C.3.: CALC

Press **CALC** (or **2nd TRACE**) to view the calculate menu.

The **CALC** menu (in **Func** mode) allows you to find values of a function, zeros of a function, min and max of a function, intersection points, derivatives and integrals.

1:value   -   evaluates a function for a given value of x.
2:zero   -   find an x –intercept.
3:minimum   -   finds a relative minimum.
4:maximum   -   finds a relative maximum.
5:intersect -   finds a point of intersection.
6:dy/dx   -   finds the derivative of a function.
7:∫f(x)dx   -   finds the value of the integral of a function.

The **CALC** menu in Parametric, Polar, and Sequence modes includes the following:

Par   **Pol**   **Seq**

C.3.1.: 1:value

The value option is used to evaluate a selected function for a particular value of x.
Press **CALC ... 1:value** to access this option.

1. Graph the function: $Y1 = -3x^3 + 12x - 3$.
Find the value for x = 1.
Press **CALC ... 1:value ... ENTER... 1** (for x=1) .... **ENTER**

  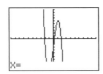 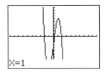 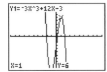

When x = 1, y = 6.

2. Graph two equations at the same time: $Y1 = x^2 - 2$ and $Y2 = x + 1$
   Find the following values: Y1(-3) and Y2(4)

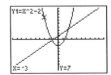   Y1(-3) = 7

How can you find the value for the other function?

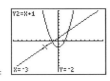

Press the down arrow key and you will get the value for Y2(-3) =

To find the value for Y2(4) just enter **4** (for x = 4), it will place it at the x = 4 and press **ENTER**

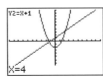 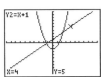   The value for Y2(4) = 5

Find Y1(-4) and Y2(0).

3. Find the value of $Y1 = \tan x$ at $x = \pi$.
   Graph **Y1** using the **ZTrig** mode

ZStandard   **ZTrig**

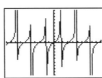 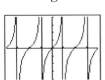 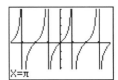 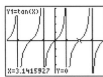

The value of Y1(π) = 0. Try other values such as Y1(1) = , Y1(-2) =   !!

40

C.3.2.: 2:zero

This option will find the zero or x-intercept of a graph. To access press **CALC ... 2:zero**.
1. Graph the function $Y1 = x^2 - 2$. Find the x-intercept to the right of 0.

CALC ... 2:zero...ENTER will give

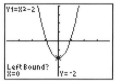

LeftBound? You need move your cursor to a point to the left of the x-intercept

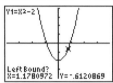   press **ENTER** and you'll see

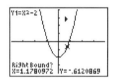   an arrow pointing to the right and RightBound? appears.

RightBound? Now move the cursor to a point to the right of the x-intercept ....**ENTER**

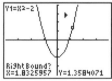

Now you'll see two arrows and Guess?

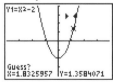   press **ENTER** again and you will get the x-intercept

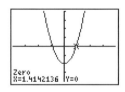   The x-intercept is x=1.4142136. Find the other x-intercept x = _____ .

2. Try the function $Y1 = x^2 - 2x - 5$. Find the x-intercepts: x = _____ and x = _____ .

3. Try the function $Y1 = e^x - 3$. Find the x-intercept: x = _____ .

C.3.3.: 3:minimum and 4:maximum

The minimum option is used to find the relative minimum of a graph.
Press **CALC...3:minimum** to access this option.

1. Graph the following function **Y1 = -(1/3)x³ + 3.5x² – 10x + 7** and find the relative minimum. Use a standard window.

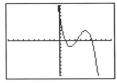

To find the minimum press **CALC ...3:minimum ... ENTER** and you'll see

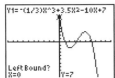   The question "LeftBound?" appears at the bottom of the screen. Use the arrow key to move the blinking cursor to the left of the relative minimum and press ENTER.

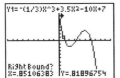   The question "RightBound?" appears at the bottom of the screen. Use the arrow key to move the blinking cursor to the right of the relative minimum

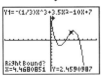   and press ENTER.

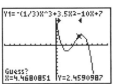   The arrows at the top of the screen indicate the boundaries between the values for which the calculator will provide the relative minimum. And the Guess? graphic appears. Just press ENTER again and you'll see the following:

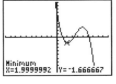   The coordinates of the minimum appear at the bottom of the screen. For our function Y1 the relative minimum is at (2, -1.67)

2. You can find the minimum over a new interval that does not contain the relative minimum of the graph. For example, find the following minimum between 2.9787 and 4.04.

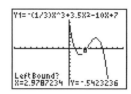 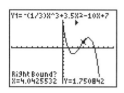 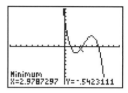

There exists a minimum which is x=2.9787 y = -.5423 since the function is increasing in this interval and does not have a turning point.

Note: To find the relative maximum of a graph use the same procedure as the one used to find the relative minimum, but you need to use **CALC … 4:maximum** .

3. Find the relative maximum for the above graph: Maximum X= _____ Y= _____ .

4. Find the relative minimum and relative maximum for the following function
$Y1 = -3x^3 - 12x^2 - 2x + 3$

C.3.4.: 5:intersect

1. Find the intersection point(s) of 2 or more graphs.
    a. To find the intersection(s) of two graphs: i.e. $y = (1/20)x^3 - 3$ and $y = -x^2 + 1$
      1) Enter the two equations in **Y=**
      2) **GRAPH**
      3) Use the **5:intersect** function to find the point of intersection(s) between the two graphs.
         a) **2nd CALC**
         b) **5:intersect**
         c) **First Curve?** Move cursor to the left of the intersection point you wish to find and press **ENTER**
         d) **Second Curve? ENTER.**
         e) The cursor will jump to the other curve. Guess? will appear then press **ENTER**
         f) **Intersection** will appear and the coordinates of the intersection point are shown: **X=-2.115 Y=-3.473**
         g) Repeat for multiple intersection points

 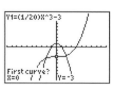 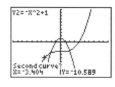 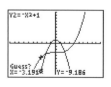 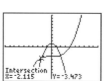

  (a) & (b)   (c)   (d)   (e)   (f)

b. Find the other intersection point of the two curves above:

Answer: Intersection point is (1.911, -2.651)

Practice: Find the intersection point of the two graphs: $y = -x^2 – 4x – 2$ and $y = x^2 + 2x -1$
Answer: Intersection points are (-.177, -1.323) and (-2.823, 1.323).

2. The **5:intersect** option can also be used to find the x-intercepts of a graph.
Find the x-intercepts of the graph: $y = x^2 + 2x + 1$. Hint: Use y = 0 for the other equation.

      a. Enter the two equations in **Y=**
      b. Graph
      c. Use the **5:intersect** function to find the point of intersection(s) between the two graphs.
           1) **2nd CALC**
           2) **5:intersect**
           3) **First Curve? ENTER**
           4) **Second Curve? ENTER.**
           5) **Guess?** will appear then press, Move cursor near the intersection in question, press **ENTER**
           6) **Intersection** will appear and the coordinates of the intersection point are shown: **X=-1.0 Y=0**
           7) Repeat for multiple intersection points

Practice: Find the intersection point(s) of the graphs: $y = x^2 + 7x + 6$.
The points you will find are also called solutions to the equation and can be used for factoring.

Answer: Intersection points are (-6, 0) and (-1, 0).

C.3.5.: 6:dy/dx and 7:∫f(x)dx will be visited at a later point.

C.4.: SHADE

There are two ways to shade a graph
                1) Using the graph style icons in the **Y=** editor.
                2) Using the **SHADE(** option.

C.4.1 Shade a graph using the graph style icons in the **Y=** editor:

Graph Styles:
\   Line   Graphs a solid line and is the default setting in **Connected** mode.
\\ Thick   Graphs a thick line.
    Above   Shades the area above the graph.
    Below   Shades the area below the graph.
-0 Path   A round cursor traces the edges of the graph and draws a path.
0 Animate A round cursor traces the edges of the graph without drawing a path.
    Dotted   A small dot represents each point; this is the default in **Dot** mode.

1. Assume you would like to shade $y \leq x$.
   a. Input your equation in **Y=** .
   b. Use the left arrow key twice to move the cursor past = sign to the graph style icon in the first column.
   c. Press **ENTER** repeatedly until you find the **Below style** \ .
   d. Press **GRAPH**.

 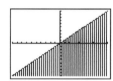

2. If you would like to shade two or more functions you will see the following:
   Graph and shade the two inequalities: $y \leq -x + 5$ and $y \geq x$:

 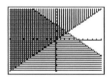

3. Shade three inequalities:
$y \leq -x + 5$
$y \geq x$
$y \leq 10x + 6$

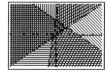

At this point you probably realize that shading three or more inequalities will result in a rather confusing graph and it is difficult to see the actual common area. For this type of problem you may want to use a different strategy, i.e. just graphing the line's curves and determining the common region manually.

Practice: Graph $y \geq x^2 + 7x + 6$.
Practice: Graph $y \leq x^3 - 8x$.
Practice: Graph $y \leq \sqrt{(8 - x^2)}$ and $y \geq \sqrt{(8 - x^2)}$.

C.4.2 Shade a graph using the **SHADE(** option in the **DRAW** menu:

With this option you can shade areas on the graph comparing two graphs.

1. Find the area where $y = x+1$ is less than $y = x^2 - 3$.
   a. Press **DRAW (or 2nd PRGM)**.
   b. Select **7:Shade(** from the menu and press **ENTER** [Tip: Clear all drawings first: 1:ClrDraw…ENTER and return]
   Shade(lowerfunc, upperfunc, Xleft, Xright) meaning it shades the area where the lowerfunc < upperfunc in the given x-interval. You can omit the Xleft and Xright if you want to cover all x's.
   c. Press **ENTER** and the graph will appear.

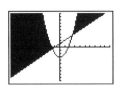

(a)　　　　(b)　　　　(c)

2. Find the area where $y = x+1$ is less than $y = x^2 - 3$ when $0 \leq x \leq 6$ only.
   a. **DRAW (or 2nd PRGM)**.
   b. **7:Shade(… ENTER**
   c. Type **Shade(x+1, x^2-3, 0, 6)**
   d. Press **ENTER** and the graph will appear.

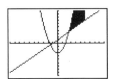

Practice: Shade the area(s) where $y = x^2 + 7x - 7$ is less than $y = 2x$. Find the x-interval where this holds true.
Practice: Shade the area(s) where $y = x^3 - 8x$ is greater than $y = 1$.

Practice Your Skills UNIT C

1. The given polynomial function can be used to estimate the number of milligrams of pain relief medication in the bloodstream at t hours after 400 mg of medication has been swallowed.

   $M(t) = 0.5 t^4 + 3.45 t^3 - 98.65 t^2 + 347.7 t$, $0 \leq t \leq 6$, t in hours

   a. Graph the above polynomial.
   b. Use the **ZoomIn** mode to magnify the $0 \leq t \leq 6$ interval.
   c. Use the **ZBox** to zoom in to the $0 \leq t \leq 6$ interval.
   d. Which zoom function will work the best?
   e. Find the relative minimum and relative maximum.
   f. Find the x-intercept(s). What do they mean in the context of the problem?
   g. During what time interval is the amount of milligram of medication more than 200mg? Estimate. (Hint: Use the **Shade(** option.)
   h. Shade the area between 2.75hrs and 3.8hrs. What is the amount of medication in the bloodstream at these time points?

2. Draw the circle $x^2 + y^2 = 60$.
   a. Find the x-intercepts?
   b. Find the relative min and relative max.
   c. Draw the graph so the circle will appear round.
   d. Find the domain and range.

3. Find the intersection points of the following functions:
   $y = [3 / (1.01x - 3)^2] + 2$ and $y = e^{(2.2 - x)}$

4. Shade the area the two functions have in common: $y \geq x - 2$ and $y \leq -x + 2$.

5. Draw a tangent line to the circle in problem 2 at x = 5.

UNIT D: Lists – StatPlot – Stat Analysis
Curve Fitting– Prob - Sim – Dist

Lists	**50-52**
StatPlot	**52-56**
Stat Analysis	**57-59**
Curve Fitting	**60-65**
Probability Computations	**65-68**
Simulations	**69-72**
Probability Distributions	**73-81**
Practice Your Skills	**82-83**

D.1.: Lists

Lists are a very powerful tool in statistics and curve fitting. Lists are part of the **STAT** feature.

D.1.1: Create Lists and Input Data

1. Using List Editor:
Let's create the following two lists: L1= 1,2,3,4,5,6,7,8,9,10 and L2= 3,2,5,4,3,6,5,1,6,2.
L1 represents the consecutive trials of rolling a die. L2 represents each outcome of rolling the die 10 times.

    a. **STAT**
    b. **4:ClrList … ENTER** and type **L1, L2, L3, … ENTER** (This will clear the lists and all data previously stored is removed)
    c. **STAT** again
    d. **1:EDIT… ENTER**
    e. Type your data in **L1**: **1 … ENTER, 2 … ENTER, …**
    f. Move to **L2** using your right arrow key.
    g. Type your data in L2.

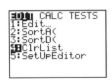   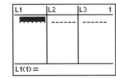 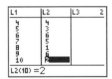

2. Create Lists on your home screen:
You can also input data into lists from your home screen. If you want to create **L3** with the values 5, 3, 2:

a. Type {5,3,2} on your home screen.
b. **STO→ L3 (2nd 3)**.
c. Check your **L3** in the **STAT** menu.

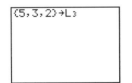 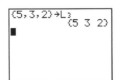

50

D.1.2.: Work with Lists
1. Basic Operations with Lists:
If you need to add L1 and 2*L2 to create L4, then do the following:
a. Place your cursor at the top of L4.
b. Type L1 + 2*L2. Notice that your equation shows up at the bottom L4= .
c. ENTER.
d. The values of L4 will appear.

2. Compute the Standard Deviation using Lists:
Let's assume you need to compute the standard deviation of the 10 outcomes of rolling the die which are in L2. Clear L3 and L4 before you start.
The formula for computing the sample standard deviation is as follows:

$$\sigma = \sqrt{\sum[(x - x(bar))^2 / (n-1)]}$$

We can use our lists to complete the lengthy computation.
a. Step 1: Find the mean x(bar) =
   You can use **LIST** (2nd STAT) …**MATH** …**5:sum(L2)** and divide by 10. The result is 3.7.
b. Create **L3 = L2 – 3.7**
c. Create **L4 = (L3)** ^{2} .
d. Now find the total of L4 from the home screen (press **QUIT** first).
   LIST … MATH … 5:sum(L4)
Which is 28.1. To find the sample standard deviation, find $\sqrt{(28.1 / 9)}$ = 1.76698… .

        (b)   **(c)**   **(d)**

D.1.3.: Sort Lists

Sometimes you need your data in ascending or descending order. Any list can be sorted in ascending order by
1. STAT
2. 2:SortA(
3. ENTER
4. L2
5. ENTER

Now check **L2**.

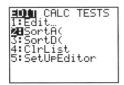  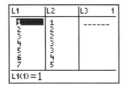

Use the same technique as above for sorting in descending order by substituting step2 with **3:SortD(**.

D.2.: StatPlot

The **StatPlot** feature allows you to create:
1. Scatter plots,
2. XY line charts,
3. Histograms,
4. Modified box-plots,
5. Box-plots,
6. Normal probability plots.

Any of these graphs use the information contained in the lists and you need to tell the calculator which lists you want to plot (default settings are always L1 and/or L2).

Clear the Y-Editor: If there is a previous entry in the Y-Editor, you need to clear all of them by pressing **Y=** ... and **CLEAR**. Use the down arrow key and **CLEAR** until all entries all cleared.

Press **StatPlot** (or 2nd Y=) to activate this menu and you will see a screen similar to this:
It depends which plots have been activated during previous operations.

Turn all Plots OFF:
You'll see that Plot1 and Plot2 are On (meaning activated). The first step in this case is to turn all plots off.

1. Select a plot by moving the cursor to i.e. **2:Plot2…On**
2. Press **ENTER**
3. Move cursor to **OFF**
4. Press **ENTER**
5. Press **QUIT**
6. Now check your **StatPlot** settings. You will see that Plot2 is turned off.

  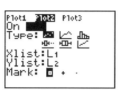 

Since you will want to work with Plot1, leave it turned ON.

D.2.1.: Scatter Plots

Create a scatter plot for our L1 and L2 from above:
1. Turn Plot1 **ON**
2. Select Plot1 and press **ENTER**
3. Use right arrow and down arrow to select the scatter plot option.
4. **ENTER**
5. Use down arrow key to select Xlist and Ylist. In our example **Xlist: L1… ENTER** and **Ylist: L2 … ENTER**
6. Choose type of marking … **ENTER**

Display the Plot
7. Press **GRAPH**
8. Use **ZOOM … 6:ZStandard and** you'll see the following:

 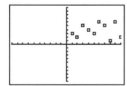   Use **ZOOM …9:ZoomStat**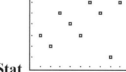

D.2.2.: XY Line Charts

To create a XY Line Chart for the same data use the following settings:

   with the graph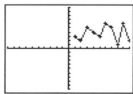

53

D.2.3.: Histograms

1. Standard Histogram:
Use the following settings for a histogram for the same data:

 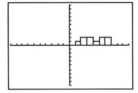

2. Customize your Histogram:
Create a histogram for a different data set. You have collected data on the height in inches of 10 female high school seniors: 62, 67, 58, 60, 66, 65, 69, 69, 74, 76. Input data in L4 and sort in ascending order. Then construct a histogram with 5 unit increments starting at 50 applying the following window settings:

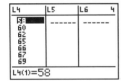 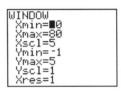 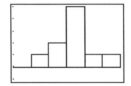

The histogram tells us that there are

Height in inches	Number of Girls
50-54	0
55-59	1
60-64	2
65-69	5
70-74	1
75-79	1

Activity: The number of calories consumed daily by a sample of 14 adults are as follows: 2340, 1200, 1500, 2900, 800, 3100, 1450, 2500, 1800, 3600, 3200, 2400, 2200, 1600.

a. Create a histogram for the given data using increments of 500 units.
b. Create a frequency table summarizing your results.
c. What would the histogram look like if the increments where changed to 1000 units?
d. Create a frequency table.

D.2.4.: Modified Box-Plots

Modified Box-Plots will show any outliers if present. Let's assume we roll the die again 10 times and our outcomes are as follows: 1,3,4,2,3,2,3,6,2,1. Use the following settings to draw a Modified Box-Plot.

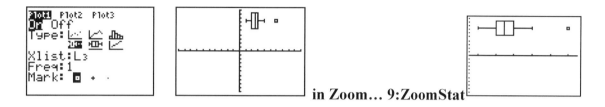

in Zoom… 9:ZoomStat

D.2.5.: Box-Plots
The settings to get a box-plot and the Box-plot itself will look as follows for the new data:

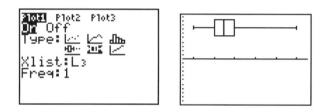

What are differences between the modified box-plot and the plain box-plot?
Could we graph both box-plots on the same screen for comparison sake?
The answer is Yes! Activate Plot1 and Plot2 as follows. Then press GRAPH and you will see this!

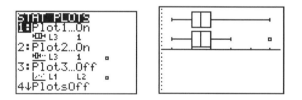

Activity:
1. Try to graph a Modified box-plot and a histogram on the same screen for the new data.
2. Try to graph a Modified box-plot and a scatter plot on the same screen for the new data.
3. Try to graph a Modified box-plot and a xy line chart on the same screen for the new data.

D.2.6.: Normal Probability Plots

A normal probability plot plots each observation in the data list versus the corresponding quantile z of the standard normal distribution. This means it will graph the value x in the list versus the standardized value of x which is $z=(x-\mu)/\sigma$. If the plotted points lie close to a straight line, then the plot indicates that the data are normal.

Normal Probability Plot for the height of 10 female high school seniors in L4:

 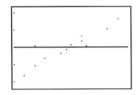 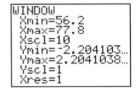

The plot resembles a fairly straight line, hence the data are normal. This result can be used for other statistical testing.

You could check this result by creating another list $L5=(x-\mu)/\sigma$ and then graph L4 vs. L5. This should produce the same graph as your normal probability plot.

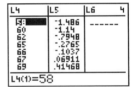 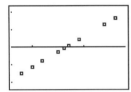 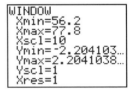

Graph both plots together:

 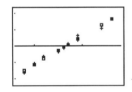

Why are some points not identical? Answer: ___

D.3.: Stat Analysis and Curve Fitting

The features described below are useful in computing statistical variables such as the mean, $\sum x$, $\sum x^2$, s_x (sample standard deviation), σ_x (population standard deviation), n (sample size), Q1 (First Quartile), Med (2nd Quartile), Q3 (Third Quartile), Min, Max for x and y variables.

The values calculated are also stored in the **VARS** menu which can be accessed by pressing **VARS ... 5:Statistics**, selected the desired variable, press **ENTER...ENTER**.

D.3.1.: 1-Var Stats
1. Compute the statistics of one variable: Let's use the data for the height in inches of 10 female high school seniors: 62, 67, 58, 60, 66, 65, 69, 69, 74, 76 and input in L1 (clear all lists before you input this data!). Compute the mean, median, sample standard deviation, Q1, Q3, min and max for this data.

   a. Input data in **L1** and **QUIT**
   b. Press **STAT**
   c. Select **CALC**
   d. **ENTER** (default is L1)
   e. **ENTER**
   f. Use down arrow key to get the rest of the statistical variables.

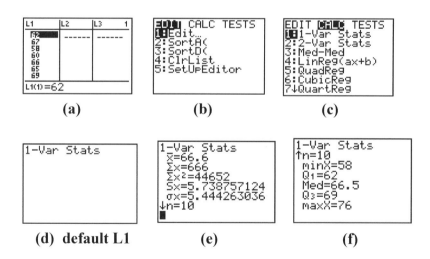

(a)   (b)   (c)

(d) default L1   (e)   (f)

Results: The mean is 66.6 in., sample std. deviation is 5.7387... in., sample size is 10, the minimum is 58 in., first quartile is 62 in., median is 66.5 in., third quartile is 69 in., and maximum is 76 in. All these values will also be stored in the VARS menu. Check it!

Activity: Compute all statistical variables for the calorie consumption data: The number of calories consumed daily by a sample of 14 adults are as follows: 2340, 1200, 1500, 2900, 800, 3100, 1450, 2500, 1800, 3600, 3200, 2400, 2200, 1600.

2. Compute the Weighted Mean:

Let's assume you are a shrimp lover and you have purchased various amounts of shrimp at different stores and markets at varying prices. In the end you would like to know how much on average you paid for one pound of jumbo shrimp. Over the last three months, you have purchased the following: 3 lbs. at $6.33 per pound, 5 lbs at $5.29 per pound, 10 lbs at $5.99 per pound, 2 lbs at $7.49 per pound and 6 lbs at $5.49 per pound. How much on average did you pay per pound?

Let x = the price of shrimp and y= the amount of shrimp. Input these values in L1 and L2. Find the weighted mean:
   a. Input data in **L1** and **L2,** then **QUIT**
   b. Press **STAT**
   c. Select **CALC** and select **1-Var Stats**
   d. **ENTER** and type **L1, L2**
   e. **ENTER**

Use down arrow key to get the rest of the statistical variables.

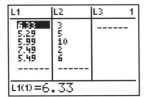   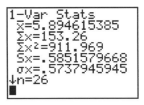

Hence, the average price per pound is $5.89.

Manual Computation:

$$\frac{3(6.33) + 5(5.29) + 10(5.99) + 2(7.49) + 6(5.49)}{3 + 5 + 10 + 2 + 6} = 5.8946\ldots$$

Activity: During the following 7 days you have been online writing as many emails as possible to inform your friends of the upcoming graduation party. The following are the lengths of time online and the numbers of letters written: Monday 5 min, 16 letters; Tuesday 15 min, 24 letters; Wednesday 8 min, 6 letters; Thursday 2 min, 3 letters; Friday 25 min, 18 letters; Saturday 1 hour, 35 letters; and Sunday 20 min, 7 letters.

D.3.2.: 2-Var Stats

Compute the statistics of two variables: The two-variable statistics feature is helpful if you are dealing with two variables at the same time, i.e. height and weight.

For example, we collected more data on the high school girls. Each girl was asked about their weight and the following was recorded:

Height (in.)	Weight (lbs.)
62	105
67	124
58	120
60	90
66	130
65	120
69	162
69	134
74	122
76	185

Compute the statistical variables for the height and the weight of the girls.
Use the same procedure as above but you need to select 2-Var Stats and indicate where you store the data, namely L1, L2.

1. Input data in **L1** and **L2** then **QUIT**
2. Press **STAT**
3. Select **CALC** and **2-Var Stats**
4. **ENTER** and type **L1, L2** (default is L1 and L2)
5. **ENTER**
6. Use down arrow key to get the rest of the statistical variables.

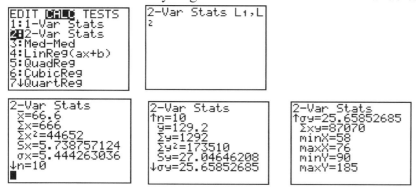

This will provide you will all the statistical variables for height and weight. Again these variables are stored in the VARS menu and you can recall them any time.

D.3.3.: Curve Fitting
Find the curve of best fit to a given set of data. This means that we need to find an equation which will model the data closely with the smallest error. You have several options:

Linear Regression	$y = ax + b$
Quadratic Regression	$y = ax^2 + bx + c$
Cubic Regression	$y = ax^3 + bx^2 + cx + d$
Quartic Regression	$y = ax^4 + bx^3 + cx^2 + dx + e$
Logarithmic Regression	$y = a + b \ln(x)$
Exponential Regression	$y = ab^x$
Power Regression	$y = ax^b$
Logistic Regression	$y = c / (1 + a*e^{(-bx)})$
Sinusoidal Regression	$y = a \sin(bx+c) + d$

The most commonly used curve fitting models are Linear, Quadratic, Cubic and Quartic, Logarithmic and Exponential Regression. The type of model you will choose depends on the type of data. Hence, it is advantageous to first graph your data using a scatter plot.

1. Linear Regression
Let's use the data Height versus Weight from the previous section. Is there a relationship between the height and weight of an individual? You can determine the line that will best fit the data by performing a linear regression using your calculator.

Height (in.)	Weight (lbs.)
62	105
67	124
58	120
60	90
66	130
65	120
69	162
69	134
74	122
76	185

Graph the data and find the best fit line using LinReg:
  a. Use a scatter plot to graph the data
  b. Input your data in **L1 and L2**
  c. **STAT**
  d. **CALC**
  e. Select **4:LinReg(ax + b)**
  f. **ENTER … ENTER**

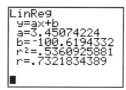

Note: If you do not get r^2 and r, turn your Diagnostic On.
  Diagnostics for SinReg is limited to y, a, and b. It does NOT provide r^2 or r.

Interpretation: Your linear regression line is as follows:
Y = 3.45 x – 100.62
Slope a = 3.45
Y-intercept of -100.62
Correlation coefficient of r = .7322
Coefficient of determination of r^2 = .5361.

Explain the slope a, y-intercept, r and r^2. What do they mean in terms of our data?
  Slope: _____
  y-intercept: _____
  r = _____
  r^2 = _____

Activity: Airline Data

American Airlines Flights Departing from Chicago

Flight	Gate-to-Gate Minutes (taxiing and flight time)	Miles
To Boston	130	868
To Dallas	140	803
To Denver	150	902
To Indianapolis	52	178
To Nashville	85	410
To New Orleans	125	838
To New York City	120	734
To Orlando	157	1006
To Toronto	86	438
To Washington, D.C.	104	613

  a. Graph the data. Is there a linear trend?
  b. Find the equation of the line of best fit using LinReg: y = _____
  c. Interpret the slope, y-intercept, r, and r^2.
  d. Find the x-intercept. What are the airplanes doing during that time?

2. Quadratic, Cubic and Quartic Regression

If the data does not follow a straight line or shows a linear trend, you may want to explore other options such as QuadReg, CubicReg and QuartReg. The model with the highest r and r^2 indicates the best possible fit. However, curve fitting is always cumbersome and tricky. Sometimes none of these models will prove to model the data very well.

Example: Gas consumption vs. Speed of car on a 250 mile trip

Speed of Car (miles) = X	Gas Consumption (in gal.) = Y
20	5.2
30	5.8
40	6.2
50	6.4
60	6.4
70	6.2
80	5.8

Find the equation of best fit:
   a. Graph the data.
       The points show the shape of a parabola, hence use **5:QuadReg**
   b. **STAT**
   c. **CALC**
   d. Select **5:QuadReg L1,L2**
   e. **ENTER … ENTER**

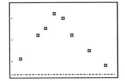 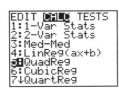 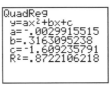

The equation of best fit is: $y = -.003x^2 + .32x - 1.61$ with $R^2 = .87$.

Graph the equation y in Y1 and compare the results. Is it a good fit?

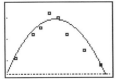

Suggestion: A quick procedure to determine which model to use is finding the differences between the y-values. Create a table and compute the first, second, third, etc. differences of y's. When the differences computed are all about the same it is an indication of the model you should use. For example, if your third differences are equal you should use a cubic model.

Speed of Car (miles) = X	Gas Consumption (in gal.) = Y	$\Delta_1 Y$	$\Delta_2 Y$	$\Delta_3 Y$
20	5.2			
30	5.8			
40	6.2			
50	6.4			
60	6.4			
70	6.2			
80	5.8			

Activity: Fit a curve to the following data.

1	-4
2	0
3	1
4	0
5	-2
6	0
7	3
8	15

a. What model should you choose?
b. Find the equation: y = _____
c. What is R^2? _____
d. Graph the data and the equation you found. Is this a good model? _____

Activity: Fit a curve to the following data.

1	4
2	1
3	.5
4	.2
5	2
6	5
7	7
8	15

a. What model should you choose?
b. Find the equation: y = _____
c. What is R^2? _____
d. Graph the data and the equation you found. Is this a good model? _____

3. Exponential and Logarithmic Regression

If your data resembles the graph of an exponential curve, use ExpReg.
In a case where your data follows a logarithmic curve, apply LnReg.

Example:
Graph the following data and decide which regression method to use.

1	.02
2	.5
3	.8
4	1.1
5	2
6	5
7	7
8	15

Activity: Find the equation of best fit for the following data.

1	-20
2	-5
3	-.8
4	1.1
5	2
6	5
7	7
8	9

D.4.: Probability Computations

Use the **MATH** menu selecting **PRB** to perform permutations, combinations, factorial and random number operations.

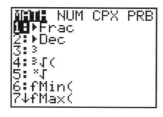 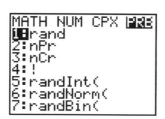

1:rand   Random-number generator
2:nPr   Number of permutations
3:nCr   Number of combinations
4:!   Factorial
5:randInt(Random-integer generator
6:randNorm(Random # from Normal distribution
7:randBin(Random # from Binomial distribution

In this section, we will visit items 2 through 4 only. Items 1 and 5 through 7 will be addressed at a later point.

D.4.1.: Permutations - Order Important

Permutations compute the number of possible selections if the order of ranking is important.

Example 1: You would like to find out how many different possibilities are there to select the president and vice-president from the 3 nominees (Al, Tom, Drake):
Answer: Your choices in the order of President, Vice-President are as follows:
(Al, Tom), (Al, Drake), (Tom, Drake), (Tom, Al), (Drake, Al), (Drake, Tom).
Hence, there are 6 possibilities to select 2 out of 3 if order is important.
You'll probably agree that it does make a difference who will be running the company, meaning who will be the president.

1. type **3** (on main screen)
2. **MATH**
3. **PRB**
4. **2:nPr**
5. **ENTER**
6. Type **2**
7. **ENTER**

The answer is 6.

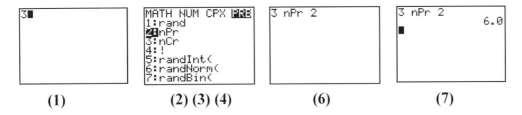

(1)　　　　　(2) (3) (4)　　　　(6)　　　　(7)

Example 2: How many ways can you select the Gold Medalist and Silver Medalist out of 8 contestants in the Down Hill Skiing event?
You will have to select all possible groupings of 2 out of 8 which is equivalent to the permutation 8P2. Use your calculator to perform this operation.

Activity 1: Race Track:
a. How many different Perfectas (pick first and second place winner in order) for a race with 7 horses?
b. How many different Quinellas (pick first, second, third and fourth place winner in order) for a race with 7 horses?

D.4.2.: Combinations – Order not important

Combinations compute the number of possible selections if the order of ranking is **not** important.

Example 1: You are taking a cruise to Cozumel Mexico with three of your friends. During one afternoon you all would like to participate in the scavenger hunt on the island of Cozumel. However, only pairs of two are allowed to participate. How many different arrangements can be made so that two pairs are formed? Participant names are Liz, Sherry, Julia, and Janene. Possible teams are: (Liz, Sherry) and (Julia, Janene), (Liz, Julia) and (Sherry, Janene), (Liz, Janene) and (Sherry, Julia) and the answer is 6.

1. Type **4** (on main screen)
2. **MATH**
3. **PRB**
4. **3:nCr**
5. **ENTER**
6. Type **2**
7. **ENTER**

The answer is 6.

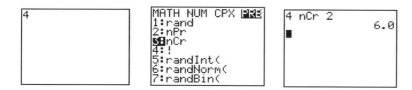

Example 2: There are 9 boys at your son's birthday party and you would like to play a game of Pictionary with teams of three. How many different teams can be formed if all the boys get along with each other? Note the order of arranging the teams does not matter, i.e. a team consisting of Alex, Pete, and Jim is the same as the team Jim, Alex, and Pete.

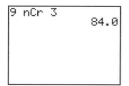

The answer is 84.

Activity 1: Race Track:
a. How many different Perfectas Boxed (pick first and second place finisher order does not matter) for a race with 7 horses?
b. How many different Trifectas Boxed (pick first, second, and third place finisher order does not matter) for a race with 7 horses?

Activity 2: Downhill Skiing:
How many ways can you select the two top finishers out of 8 contestants in the Down Hill Skiing event?

D.4.3.: Factorial !
Factorial computes all possible arrangements of all candidates present. Only integers or multiples of .5 can be used.

.n! = n * (n-1) * (n-2) * (n-3) * ... * 3 * 2 *1
4! = 4 * 3 * 2 * 1 = 24

Example 1: How many different ways can 6 figure skaters place after one round of competition? The first contestant could finish in place 1 or place 2 or ... place 6. Hence, there are 6 possible outcomes. The second contestant could place in all places except the one taken by the first contestant. Hence, there are 5 possible outcomes. The third contestant could place in all places except the ones taken by the first and second contestant. Hence, there are 4 possible outcomes. Do you see a pattern? How many outcomes are possible for the fourth, fifth and sixth contestant? 3, 2, 1

So, we now have: 6 * 5 * 4 * 3 * 2 * 1 = 6! = 720 possibilities for all contestants to place.

1. Type **6** (on main screen)
2. **MATH**
3. **PRB**
4. **4:!**
5. **ENTER**
6. **ENTER**

The answer is 720.

Example 2: You and your partner are going to a movie with 2 other couples and you are wondering how many different seating arrangements are possible if each couple sat together. (Couple 1, Couple 2, Couple 3).

3! = 6

Activity 1: Race Track: How many different ways can 7 horses finish in one race?

Activity 2: At a family gathering you would like to line up your 12 relatives in a row. How many ways can this be done?

D.5.: Simulations

The TI-83 calculator allows us to simulate the following numbers using the **MATH** and **PRB** menu:
  a. **1:rand**   random numbers from 0 to 1, inclusively,
  b. **5:randInt(**   random Integers
  c. **6:randNorm(**   Random # from Normal Distribution
  d. **7:randBin(**   Random # from Binomial Distribution

D.5.1.: Random-Number Simulation ≥ 0 and ≤ 1
1. Generate random numbers ≥ 0 and ≤ 1:
Rand(generates and returns one or more random numbers ≥ 0 and ≤ 1 and you only need to specify the number of trials (an integer > 1) you wish to generate.

Let's assume you would like to generate 5 numbers between 0 and 1, inclusively.
  a. **MATH**
  b. **PRB**
  c. **rand(5)**
  d. **ENTER**

Note: Every trial will create different numbers since they are randomly generated.

2. Generate random numbers beyond the range of 0 to 1:
  a. **MATH**
  b. **PRB**
  c. **Type rand*4**
  d. **ENTER**

3. Generate 3 random numbers beyond the range of 0 to 1:
  a. Type **rand(3)*4**
  b. **ENTER**

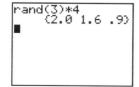

4. Store random numbers in Lists:
You can also store the numbers generated in your list editor. Generate 5 random numbers and store in L1:
  a. **MATH**
  b. **PRB**
  c. **rand(5) →L1**
  d. **ENTER**

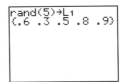   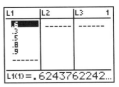

Check your List!

D.5.2.: Simulate Random-Integer
randInt(will generate and display one or more random integers within a range specified by lower and upper integer bounds: **randInt(lower, upper, number of trials)**

1. Generate one random integer:
Let's say you would like to generate random integers between 10 and 20, inclusively.
  a. **MATH**
  b. **PRB**
  c. **5:randInt(**
  d. type **10, 20)**
  e. **ENTER**

2. Generate more than one random integer:
Let's say you would like to generate 5 random integers between 10 and 20, inclusively, and store in L2.
  a. **MATH**
  b. **PRB**
  c. **5:randInt(**
  d. type **10, 20,5) →L2**
  e. **ENTER**

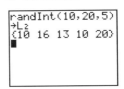

Store in L2

D.5.3.: Simulate Random # from Normal Distribution

randNorm(generates and displays a random real number from a specified Normal distribution with mean μ and standard deviation σ. Most generated values will be in the interval [μ - 3σ, μ + 3σ].

1. Generate one value of a normal distribution:
Let's say you would like to generate a random real value from a normal distribution with μ = 10 and σ = 2.
a. MATH
b. PRB
c. 6:randNorm(
d. type **10, 2)**
e. ENTER

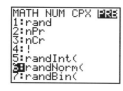 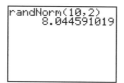

2. Generate more than one value of a normal distribution:
Let's say you would like to generate 5 random real values from a normal distribution with μ = 10 and σ = 2 and store in L1.
a. MATH
b. PRB
c. 6:randNorm(
d. type **10, 20,5)** →L1
e. ENTER

 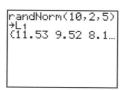   check your List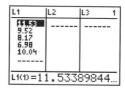

Activity: You performed a survey asking your students how many candy bars they eat during one week. You computed the mean and standard deviation of your collected data and reached the following results: μ = 12 and σ = 3. Generate a sample data with your calculator which would simulate asking 15 more students and store in L1.

D.5.4.: Simulate Random # from Binomial Distribution

randBin(generates and displays one or more random real numbers from a specified Binomial distribution with n = number of trials, p = probability of success (must be ≥ 0 and ≤ 1) and the number of simulations desired.

randBin(number of trials, prob. of success, number of desired simulations)

1. Generate one value of a Binomial distribution:
Let's say you would like to generate a random real value from a Binomial distribution with n = 8 and p = .5. For example, this could represent the number of tails you would get if you flip a fair coin 8 times with probability of tails equal to .5.
a. MATH
b. PRB
c. 7:randBin(
d. type **8, .5)**
e. ENTER

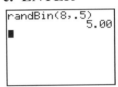

This result tells you that you had 5 tails in 8 flips of the fair coin.

2. Generate more than one value of a binomial distribution:
Let's say you would like to generate 6 random real values from a binomial distribution with n = 8 and p = .50. This could represent the outcomes you observe when flipping a fair coin 8 times over and over.
a. MATH
b. PRB
c. 7:randBin(
d. type **8, .5,6)**
e. ENTER

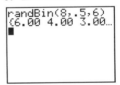

This result reveals the following:
After the first 8 flips of the coin you observed 6 tails.
After the second 8 flips of the coin you observed 4 tails.
After the third 8 flips of the coin you observed 3 tails.
After the fourth 8 flips of the coin you observed ….
Can you fill in the rest?

Activity 1: Could you simulate rolling a fair die? Let's assume you would like to know how many 6's could you expect in 10 rolls of a fair die?

Activity 2: Your chance of getting online at the first try is .4. How many times could you expect to get online right away if you log on 20 times?

D.6.: Probability Distributions

Use the **DISTR menu** to work with Distribution Functions.

DISTR
1:normalpdf(
2:normalcdf(
3:invNorm(
4:tpdf(
5:tcdf(
6: χ^2 pdf(
7: χ^2 cdf(
8:Fpdf(
9:Fcdf(
0:binompdf(
A:binomcdf(
B:poissonpdf(
C:poissoncdf(
D:geometpdf(
E:geometcdf(

DRAW
1:ShadeNorm(
2:Shade_t(
3:Shadeχ^2(
4:ShadeF(

DISTR or (2nd VARS) gives you the following screen on your calculator:

D.6.1.: Binomial Distributions
You will need to scroll down to **0:binompdf(** .
This feature computes the probability for a binomial distribution at a specified x with given number of trials = n, and probability of success = p.
1. Binompdf(number of trials, probability, x)

Example 1: What is the probability of getting 3 6's when rolling a fair die 5 times?
n = 5, p = 1/6, x = 3
    a. **DISTR**
    b. Select **0:binompdf(**
    c. **ENTER**
    d. Type **5,1/6,{3})**

e. **ENTER**

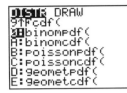 

The answer is .03.

Example 2: You take a multiple choice exam that consists of 10 questions. Each question has four possible answers, only one of which is correct. To complete the quiz, you randomly guess the answer to each question. The random variable x represents the number of correct answers.
  a. Identify p = probability of success (correct answer)
  b. Find the probability that you will guess 4 answers correctly.
  c. Find the probability that you will guess 4 or 5 answers correctly.

Answer:
a. p = 1/4
b. n = 10, p = 1/4, x = 4, Binompdf(10,1/4,{4})
  1. **DISTR**
  2. Select **0:binompdf(**
  3. **ENTER**
  4. Type **10,1/4,{4})**
  5. **ENTER**

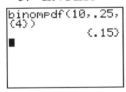

The answer is .15.

c.
  1. **DISTR**
  2. Select **0:binompdf(**
  3. **ENTER**
  4. Type **10,1/4,{4,5})**
  5. **ENTER**

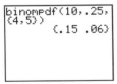

Answer: The probability of getting 4 correct is .15 and 5 correct is .06.

Activity 1: You take a multiple choice exam that consists of 10 questions. Each question has five possible answers, only one of which is correct. To complete the quiz, you randomly guess the answer to each question. The random variable x represents the number of correct answers.
   a. Find the probability that you will guess 2 answers correctly.
   b. Find the probability that you will guess 7 or 8 answers correctly.

Activity 2: Forty-five percent of working adults spend less than 25 minutes commuting to their jobs. If you randomly select six working adults, what is the probability that exactly 3 of them spend less than 25 minutes commuting to work?

2. Binomcdf(number of trials, probability, x)
The binomcdf computes the cumulative probability at x for a binomial distribution.

Example 1: What is the probability of getting 2 or less 6's when rolling a fair die 5 times?
Answer: The problem at hand now asks us to compute the probability of getting two 6's, one 6 and no 6 at all, and then add all these probabilities together. The **binomcdf(** will do exactly that, it will accumulate all the values of x and below. Our values are still n = 5, p = 1/6, x = 2.

    a. DISTR
    b. Select **A:binomcdf(**
    c. ENTER
    d. Type **5,1/6,2)**
    e. ENTER

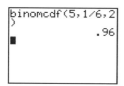

Example 2: Use the multiple choice exam problem above: n = 10, p = 1/4. What is the probability that a student will guess 5 or less correctly?
Answer: Input **Binomcdf(10,1/4,5)** which equals .98.

Activity 1: What is the probability of getting 5 or more 6's when rolling a fair die 5 times?

Activity 2: Multiple Choice Exam: What is the probability that a student will guess 8 or more questions correctly given that there are 4 possible answers only one of which is correct? (n=10)

D.6.2.: Poisson Distributions

The Poisson distribution computes the probability of an experiment where you are given the average number of occurrences over time, i.e. on a given intersection 4 accidents occur daily. The underlying assumptions are that the probability of an event occurring remains the same and that the occurrences are independent of each other.

The formula for the Poisson Distribution is as follows:

$$P(x) = (\mu^x e^{-\mu})/ x!$$ Poissonpdf(μ, x), where x is a specified number.

1. poissonpdf(

Computes the probability for a specified value of x.

Example 1: We know that on the average 4 accidents occur at the intersection of Rt. 64 and Randall Road during any given week.
a. What is the probability that only 2 accidents occur in any given week?
b. What is the probability that 7 or 8 accidents occur in any given week?

Answer: to a.
1) **DISTR**
2) **B:poissonpdf(**
3) **ENTER**
4) type **4,2)**
5) **ENTER**

 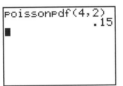

The answer is .15.

Answer: to b. Use poissonpdf(4,{7,8}) which is equal to

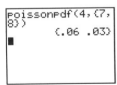

The probability of 7 business failures is .06 and the probability of 8 business failures is .03.

Example 2: The Tacoma Narrows Bridge in Washington State was the first suspension bridge built in the US. The average occupancy of vehicles traveling across that bridge during a five day period is 1.6. What is the probability that a vehicle with five people will travel across the bridge?
Answer: poissonpdf(1.6,5) is equal to .02.

Activity: The mean number of business failures per hour in the US in recent years was about 8. What is the probability that exactly 5 or 6 businesses will fail in any given hour?

2. poissoncdf(

Computes the cumulative probability of a value x and less than that, i.e. at least 2 results in adding the probabilities of 2, 1, or 0.

Example 1: We know that on the average 4 accidents occur at the intersection of Rt. 64 and Randall Road during any given week.
a. What is the probability that at most two accidents occur in any given week?
b. What is the probability that 7 or less, 8 or less and 9 or less accidents occur in any given week?

Answer to a.   Use **poissoncdf(4,2)**
1) **DISTR**
2) **C:poissoncdf(**
3) **ENTER**
4) type **4,2)**
5) **ENTER**

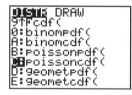 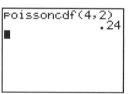

Answer to b.   Use **poissoncdf(4,{7,8,9})**

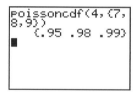

The probabilities are as follows: 7 or less accidents occurring is .95, 8 or less accidents occurring is .98, and 9 or less accidents occurring is .99.

Example 2: Use the business failure example from above. What is the probability that at least 4 businesses will fail in any given hour.?
Answer: Use **poissoncdf(8,4)** which is .10

Activity 1: The mean number of oil tankers at a port city is 9 per day. The port has facilities to handle up to 14 tankers in a day. What is the probability that too many tankers arrive on any given day?

Activity 2: The mean number of strokes per hole for golfer Tiger Woods is about 3.9. What is the probability that he will need more than 5 strokes for any given hole?

D.6.3.: Geometric Distributions

A geometric distribution is used in cases where you perform an experiment until you reach success, i.e. how many bottle caps do you need to purchase in a bottle cap contest to win? Use **geometpdf(** and **geometcdf(**.

$P(x) = p(q)^{(x-1)}$, p is the probability of success, q is the probability of failure, x is the number of trials

1. geometpdf(p,x)

Finds the probability at x, the number of the trial at which the first success occurred.

Example 1: You know that in a bottle cap contest every 3rd bottle cap has a winning prize. What is the probability that you will get the winning bottle cap at the 8th try?

Answer: p = 1/3, x = 8, so use **geometpdf(1/3, 8)**
a. **DISTR**
b. **D:geometpdf(**
c. **ENTER**
d. type **1/3,8)**
e. **ENTER**

```
geometpdf(1/3,8)
              .02
```

The answer is .02.

Example 2: What is the probability that a family will have their first boy with the 5th child born? (Probability of a boy is .49)
Answer: p = .49, 5
Geometpdf(.49,5) which is equal to .03.

Activity: The probability that a student passes the written test for a pilot's license is .75. What is the probability that a student will fail the test on the first two trials and pass it on the third trial?

2. geometcdf(p,x)

Finds the cumulative probability at x, the number of the trial at which the first success occurred.

Example: What is the probability that you will make a sale with at most 4 phone calls, given the probability of success (making a sale on the phone) is .12?

Answer: Use **geometcdf(.12,4)** which is .40.

Activity: The probability that a student passes the written test for a pilot's license is .75. What is the probability that a student will fail the test on the first trial and pass it on the second trial? This is equivalent to saying that the student will fail the test at most once.

D.6.4.: Normal Distributions

Computes the probability for the density function of a normal distribution with a mean μ, standard deviation σ and a specified value x. Draw a normal curve (bell curve) and place the mean in the middle.

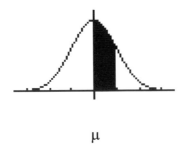

μ

1. Normalpdf(x, μ, σ) is not really useful since the probability at any given point on the normal distribution cannot be computed.

2. Normalcdf(lowerbound, upperbound, μ, σ) is much more practical

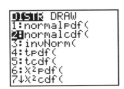

Example 1: The average cholesterol level of American men is normally distributed with a mean of 210 and a standard deviation of 22. What is the probability that a randomly selected male has a cholesterol level of less than 180?

Answer: Use **Normalcdf(-1E99,180,210,22)** which will find the cumulative probability of all values less than 180. We need to start at -∞ and include the entire interval up to 180.

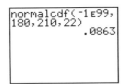

Example 2: For the above cholesterol example, what is the probability that a randomly chosen male has a cholesterol level between 190 and 220?

Answer: Use **Normalcdf(190,220,210,22)** which is .4936

Example 3: Compute the probability that x is between 0 and 1 for a standard normal distribution N(0,1).
Answer: Use **Normalcdf(0,1,0,1)** which is .3413.

3. Draw a Normal curve and Shade the Area

Use **DISTR** and **DRAW** and **ShadeNorm(** menu to shade the area covered under the normal curve. Before using this feature set the window appropriately.

Example 1: To shade the standard normal curve N(0,1) from 0 to 1 perform the following:

    a. **DISTR**
    b. **DRAW**
    c. **1:ShadeNorm(**
    d. **ENTER**
    e. Type **0,1,0,1)**
    f. **ENTER**

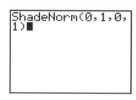 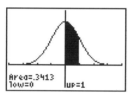 Window used:

Example 2: Shade the normal distribution with mean 210 and standard deviation 22 between 190 and 220.

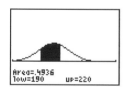 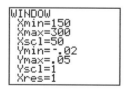

Activity 1: Shade N(5,1) between 5 and 7.

Activity 2: Shade N(0,1) between -2.5 and 2.5.

Practice your Skills UNIT D

1. Use the data on page 84 to complete the following problems (see data set p. 84):
   a. Input Section 1 Grade in L1 and Section 2 Grade in L2.
   b. Compute the standard deviation of L1 using lists.
   c. Compute 1-Var Statistic for L1.
   d. Compute 2-Var Statistic for L1 and L2.
   e. Draw a scatter plot for L1.
   f. Draw a xy line chart for L1.
   g. Draw a histogram for L1.
   h. Draw a histogram for L2.
   i. Draw a boxplot and a modified boxplot for L1.
   j. Draw a boxplot and a modified boxplot for L2.
   k. Draw a normal probability plot for L1.

Combinatorics:
2. How many handshakes?
   Determine how many handshakes are possible if everyone shakes each others hands at a business meeting.

# of people	1	2	3	4	5	6
# of handshakes						

   Use combination to find the number of possible handshakes.

3. How many different ways can you select the lottery numbers?
4. How many different ways can you select the president, vice president and secretary from a pool of 7 qualified candidates?
5. How many different ways can 6 people sit in a circle?

Simulations:
6. Simulate 10 random numbers from 0 to 5 inclusively.
7. Simulate the height of 10 individuals.
8. Simulate the number of boys if a family has 6 children. (Use randBin)

Distributions:
9. Find the probability of having at most 3 girls among 5 children.
10. On average 3 people wait in line at the supermarket check out. What is the probability that there will be 5 people waiting in line when you arrive?
11. There is a 3 in 10 chance of getting the winning gold sticker when purchasing a CD. What is the probability that you will get the winning gold sticker after purchasing 20 CD's?
12. The average number of Cokes purchased per month is normally distributed with a mean of 14 and a standard deviation of 3. What is the probability that a randomly selected individual purchased between 12 and 18 Cokes during the month of June?
13. Draw and shade the above distribution.

Curve Fitting:
14. Quadratic Regression
    a. Alternate approach to finding a model for problem 2:
    Determine how many handshakes are possible if everyone shakes each others hands at a business meeting.

# of people	1	2	3	4	5	6	
# of handshakes	0	1	3	6	10	15	
Double Data	0	2	6	12	20	30	
Factors		1 2	2 3	3 4	4 5	5 6	(n-1) n

$$Y = \frac{n(n-1)}{2} = .5n^2 - .5n$$

   b. Use the standard approach, meaning Quadratic Regression, to find the equation of best fit.

15. DO THE WAVE!
    Clock the time it takes for 2, 4, 6, … people to do the wave.
    a. Find the model of best fit.
    b. Predict how long will it take for 100, 5000, 10000 people to do the wave?
    c. Predict how long it will take to do the wave at Wrigley Field (assume large attendance)?

# of people	0	2	4	6	8	10
Time						

16. M & M Problem!
    Start with 30 M&M's in a cup. Empty the cup onto a piece of paper, count the number of M&M's up and place those back into the cup. Empty the cup onto piece of paper again, count the number of M&M's which are up and place those back into the cup. Continue until you have none left.

# of M&M's	30									
# of M&M's up										

Find a mathematical model which you could use to predict the number of M&M's at any time.

17. Investment: Let's say you have $2000 to invest. If you find a bank that will pay you 8% interest compounded monthly for 10 years, how much money will you have after 1 year, 2 years, … 10 years. Find a mathematical model which would predict the amount of your investment at any time during the investment period, i.e. 42 months.

Student #	Section 1 Grade	Section 2 Grade
1	90.0	99.0
2	89.0	98.0
3	66.0	100.0
4	77.0	100.0
5	99.9	99.9
6	66.0	100.0
7	57.0	98.0
8	44.0	44.0
9	66.0	99.0
10	78.0	33.0
11	88.0	88.0
12	99.0	99.0
13	97.0	97.0
14	100.0	100.0
15	99.0	95.0
16	99.0	59.0
17	97.0	47.9
18	88.0	100.0
19	92.0	100.0
20	93.0	28.2

UNIT E: Matrix – Matrix Op - Simult Equations – Other

Matrices   86-90
Matrix Operations   91-94
Solving Simultaneous Equations   95-100
More Matrix Operations   101-103
Practice Your Skills   104

E.1.: Matrix

What is a Matrix? A matrix is a two-dimensional array. The matrix editor lets you display, define, or edit a matrix. The TI-83 can store up to 10 matrices, [A] through [J]. Every matrix may have up to 99 rows or columns if memory is available, but you can only store real numbers.

E.1.1.: Select a Matrix
1. Press **MATRIX** (or 2nd x^{-1})
2. Use arrow to select **EDIT**
   The dimensions of any previously defined matrix or matrices are displayed.
3. Select the matrix, let's say **[B]**
4. **ENTER** (the MATRIX EDIT menu will appear)
5. Type **5 ENTER 2** to define a 5 x 2 matrix (5 rows, 2 columns)
6. To change the dimensions just arrow back up and left to make changes.

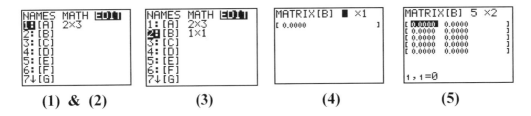

(1) & (2)   (3)   (4)   (5)

E.1.2.: Edit Matrix Elements
You want to input the following matrix in B

$$\begin{bmatrix} 2 & 3 \\ -1 & 2 \\ 4 & -4 \\ -3 & -2 \\ -5 & 6 \end{bmatrix}$$

1. **MATRIX**
2. **EDIT**
3. Select matrix **[B]**
4. **ENTER** (the MATRIX EDIT menu will appear)
5. Type **5 ENTER 2 ENTER** (Enter or accept the dimension (5 x 2))
6. Press your arrow keys to move to the element you want to change or input.
7. Input **2** into element (1,1) **ENTER**
8. Input **3** into element (1,2) **ENTER**
9. Input **-1** into element (2,1) **ENTER**
10. Etc. ... until all elements are stored

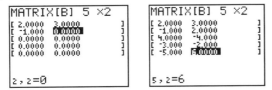

E.1.3.: Viewing a Matrix and Displaying Matrix Elements
View a matrix you have edited and check all its elements. For example, you would like to view the matrix you have just input above (matrix B):

1. **MATRIX**
2. **2:[B]** (Select desired matrix)
3. **ENTER** and **ENTER**

Matrix B will appear on the screen and you can check the elements.

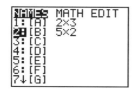 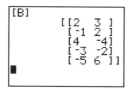

E.1.4.: Enter a Matrix from the Home Screen
To input the matrix B from above type the following in the home screen:
1. **[[2,3][-1,2][4,4][-3,-2][-5,6]]**   **Note:** To input **[** use: **[** or **2nd x**
2. **ENTER**   To input **]** use: **]** or **2nd —**

and you will see

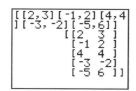

E.1.5.: Copy One Matrix to Another
If you want to copy Matrix B to Matrix A perform the following:
1. **MATRIX**
2. Select **2:[B]**
3. **ENTER**
4. **STO→**
5. **MATRIX**
6. Select **1:[A]**
7. **ENTER** and **ENTER**

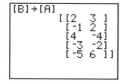 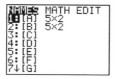

E.1.6.: Accessing a Matrix Element

From the home screen you can store a value to, or recall a value from, a matrix element. The only condition it needs to satisfy is that the element must be within the currently defined matrix dimension.

a. Store a Value: You would like to change the element (3,1) to -9 in matrix B from above
1. Type **-9 →**
2. **MATRIX**
3. Select **2:[B] ENTER**
4. Type **(3,1) :**
5. **MATRIX**, select **2:[B]**
6. **ENTER**

You will see the new matrix [B]

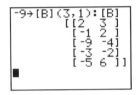

b. Recall a value: Check what the element (5,2) is
1. **MATRIX**
2. Select **2:[B]**
3. Type **(5,2)**
4. **ENTER**

And you will the element (5,2) is 6

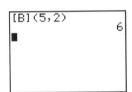

Activity 1: Input the following matrices:

$$A = \begin{bmatrix} 1 & 0 \\ -2 & 3 \end{bmatrix} \qquad B = \begin{bmatrix} 1 & 0 & 1 \\ 2 & 0 & 0 \\ 1 & 1 & -1 \\ -1 & 2 & 5 \end{bmatrix}$$

Activity 2: Check the elements of matrix A and B.
Activity 3: Change element (2,2) to 5 in matrix A, and change element (4,3) to 0 in matrix B.
Activity 4: Copy matrix A to matrix D.
Activity 5: Check the value of element (3,3) of matrix B.

E.1.7.: Row Swap: rowSwap(matrix, rowA, rowB)
This allows you to manually manipulate rows, i.e. swap rows.
Let's say you have the following matrix and you need to swap row A and row B.

$$\begin{bmatrix} 2 & 5 & 7 \\ 8 & 9 & 1 \\ 2 & 6 & 5 \end{bmatrix}$$

1. Input above matrix as **[F]**
2. **MATRIX** and **MATH**
3. select **C:rowSwap(**
4. **ENTER**
5. **MATRIX**
6. select **6:[F]**
7. **ENTER**
8. type **,1,2)**
9. **ENTER**

 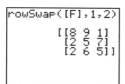

E.1.8.: Row Addition: row+(matrix, rowA,rowB)
This allows you to manually add two rows. .
Let's say you have the following matrix (store as [E]) and you need to add row A and row B. It will store the results in row B.

$$\begin{bmatrix} 1 & 5 & 7 \\ 6 & 9 & 1 \\ 3 & 6 & 5 \end{bmatrix}$$

1. Input above matrix as **[E]**
2. **MATRIX** and **MATH**
3. select **D:row+(**
4. **ENTER**
5. **MATRIX**
6. select **5:[E]**
7. **ENTER**
8. type **,1,2)**
9. **ENTER**

 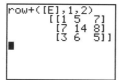

E.1.9.: Row Multiplication: *row(value,matrix,row)

This operation multiplies a particular row of a matrix by a value specified and stores the result in that row. Let's assume you need to multiply row C of matrix [E] by -2.

$$\begin{bmatrix} 1 & 5 & 7 \\ 6 & 9 & 1 \\ 3 & 6 & 5 \end{bmatrix}$$

1. Input above matrix as **[E]**
2. **MATRIX** and **MATH**
3. select **E:row+(**
4. **ENTER**
5. type **-2,**
6. **MATRIX**
7. select **5:[E]**
8. **ENTER**
9. type **,3)**
10. **ENTER**

 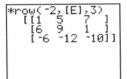

E.1.10.: Row multiplication and addition: *row+(value,matrix,rowA,rowB)

You can also multiply and add two rows at the same time. It will multiply rowA of the matrix by the indicated value, adds it to rowB, and stores the result in rowB.

Input the matrix below in [G] and multiply rowA by 2 and add to rowB.

$$\begin{bmatrix} 1 & 1 & 3 \\ -2 & 2 & -1 \\ 1 & 3 & 4 \end{bmatrix}$$

1. Input above matrix as **[G]**
2. **MATRIX** and **MATH**
3. select **:*row+(**
4. **ENTER**
5. type **2,**
6. **MATRIX**
7. select **5:[G]**
8. **ENTER**
9. type **,1,2)**
10. **ENTER**

 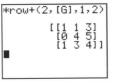

E.2.: Matrix Operations

E.2.1.: Add and Subtract Matrices
You can only add and subtract matrices of the same dimensions. It will add the elements of each corresponding matrix, i.e. element (1,1) of matrix A is added to element (1,1) of matrix B, and hence we have 2 + 0 = 2

$$A = \begin{bmatrix} 2 & 3 \\ -1 & 2 \\ 4 & -4 \end{bmatrix} \quad B = \begin{bmatrix} 0 & -1 \\ 2 & 4 \\ 8 & -2 \end{bmatrix}$$

Add A + B
1. Input matrix A and B (Note: they are both of dimension (3 x 2))
2. **MATRIX 1:[A]**
3. **ENTER**
4. Type **+**
5. **MATRIX 2: [B]**
6. **ENTER**
7. **ENTER**

```
[A]+[B]
       [[2   2 ]
        [1   6 ]
        [12  -6]]
```

Subtract A - B will give you

```
[A]-[B]
       [[2   4 ]
        [-3  -2]
        [-4  -2]]
```

E.2.2.: Multiply Matrices

a. Scalar Multiplication of a matrix
   For example you would like to get 3 * A
   1. Type **3 ***
   2. **MATRIX**
   3. select **1:[A]**
   4. **ENTER** and **ENTER**

You will get

```
3*[A]
       [[6    9  ]
        [-3   6  ]
        [12   -12]]
```

b. Multiply two Matrices

To multiply two matrices together, i.e. matrix A * matrix B, you must have a match of the column dimension of matrix A and row dimension of matrix B.

(3 x 2) * (2 x 3) will work

(3 x 2) * (3 x 2) will not work

Input a new matrix for B: $\begin{bmatrix} 1 & 2 & 4 \\ -1 & 3 & -2 \end{bmatrix}$

Multiply A * B
1. MARTIX
2. Select **1:[A]**
3. **ENTER**
4. Type *
5. **MATRIX**
6. Select **2:[B]**
7. **ENTER** and **ENTER**

You will see the following:

```
[A]*[B]
    [[-1  13   2 ]
     [-3   4  -8]
     [ 8  -4  24]]
```

E.2.3.: Negate a Matrix

Negating a matrix will result in a matrix in which the sign of every element is changed (reversed). For example, you would like to find − matrix A

1. Type (−)
2. **MATRIX**
3. select **1:[A]**
4. **ENTER**

```
-[A]
    [[-2  -3]
     [ 1  -2]
     [-4   4]]
```

E.2.4.: Absolute Value of a Matrix

This function returns a matrix containing the absolute value of each element of the matrix. To find the absolute value of matrix A use **abs(matrix):**

1. **MATH**
2. select **NUM**
3. select **1:abs(**
4. **ENTER**
5. **MATRIX**
6. Select **1:[A]**

7. **ENTER** and type **)**
8. **ENTER**

and you will get the following matrix.

```
abs([A])
        [[2 3]
         [1 2]
         [4 4]]
```

E.2.5.: Round Values of a Matrix

This option lets you round the elements of a matrix to a desired number of decimal places (≤9). If you would like to round all the elements to 2 decimal places perform the following:

Round(matrix[,#decimals])

Input Matrix C = $\begin{bmatrix} 2.5435 & -3.2122 \\ 1.5567 & 2.1111 \end{bmatrix}$

Round Matrix C to two decimal places:
1. **MATH**
2. select **NUM**
3. select **2:round(**
4. **ENTER**
5. **MATRIX**
6. Select **1:[C]**
7. **ENTER**
8. Type **,2)**
9. **ENTER**

```
[C]
[[2.5435 -3.212…
 [1.5567 2.1111…
round([C],2)
[[2.54 -3.21]
 [1.56 2.11 ]]
```

E.2.6.: Inverse of a Matrix

The matrix must be a square matrix in order to be able to find an inverse and the determinant cannot equal zero. User **matrix** $^{-1}$.

Find the inverse of matrix C (set your mode to two decimal places)
1. **MATRIX**
2. Select **3:[C]**
3. **ENTER**
4. select x^{-1}
5. **ENTER**

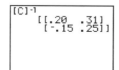

The inverse is as follows:

E.2.7.: Powers of a Matrix

To raise a matrix to a power you must first of all have a square matrix. You can take matrices to powers of 0, 1, 2, 3, ... up to 255. Use **matrix^2, matrix^3, matrix^power.**

Take matrix C to the 2nd power:
1. **MATRIX**
2. select **3:[C]**
3. **ENTER**
4. press **x^2**
5. **ENTER**

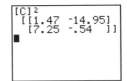

E.2.8.: Relational Operations

This function allows you to compare and test two matrices. Again you can only compare two matrices of identical dimensions.

Enter matrix D = $\begin{bmatrix} 0 & 1 & -1 \\ -1 & 2 & 5 \end{bmatrix}$ and E = $\begin{bmatrix} 5 & 1 & 2 \\ -1 & 2 & 5 \end{bmatrix}$

Matrix A and D are of the same dimension:

Let us test **matrix D = matrix E.** This will compare each element of D to E. It will return 1 if every comparison is true; it returns 0 if any comparison is false.
1. **MATRIX**
2. select **4:[D]**
3. **ENTER**
4. **TEST** (or 2nd MATH) and select **1:=**
5. **ENTER**
6. **MATRIX**
7. select **5:[E]**
8. **ENTER** and **ENTER**

 Hence, at least one comparison is false.

You can test **matrix A ≠ matrix B.** It will return a 1 if at least one comparison is false and return a 0 if no comparison is false.

Activity: Test matrix A ≠ matrix D

E.3.: Solving Simultaneous Equations

Methods used to solve two or more linear equations with two or more unknowns involve the applications of matrices. In this section, we will illustrate several methods which use matrices.

Example: Solve the following system of equations:
$2x + 3y - 4z = 2$
$-3x + 7y + 2z = 5$
$6x + 2y - z = 0$

Enter the elements of the augmented matrix into matrix [A] with dimensions 3 x 4:

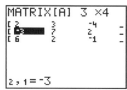

E.3.1.: Solving Simultaneous Equations using Row-Echelon Form and Back-Substitution.
Use **ref(** to transform an augmented matrix into row-echelon form. You can then use back-substitution to solve the equations for the variables. Ref will produce a matrix with 1's on the diagonal and 0's in the lower triangle.

1. Enter the elements of the augmented matrix (Use the above matrix **[A]**)
2. Select **MATRIX**
3. **MATH**
4. select **A:ref(**
5. **ENTER**
6. **MATRIX**
7. select **1:A** (or corresponding matrix with data)
8. **ENTER**
9. type **)**
10. **ENTER**

 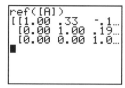

The following matrix with 1's along the diagonal and zeros below should appear on the screen (scroll to the right to see the rest):

$$\begin{bmatrix} 1 & .33 & -.17 & 0 \\ 0 & 1 & .19 & .63 \\ 0 & 0 & 1 & -.13 \end{bmatrix}$$ which implies $x + .33y - .17z = 0$
$y + .19z = .63$
$z = -.13$

Using back-substitution you will get: $z = -.13$, $y = .65$ and $x = -.24$.

E.3.2.: Solving Simultaneous Equations using Reduced Row-Echelon Form.

Use **rref(** to transform an augmented matrix into reduced row-echelon form. This form will reduce the matrix to 1's along the diagonal and 0's in the upper and lower triangle. Hence, each row will reveal the solution to all of the variables.

Example: Use the same system of equations as above and find the solutions using the rref(function.
1. Enter the elements of the augmented matrix (Use the above matrix **[A]**)
2. Select **MATRIX**
3. **MATH**
4. select **B:rref(**
5. **ENTER**
6. **MATRIX**
7. select **1:A** (or corresponding matrix with data)
8. **ENTER**
9. type **)**
10.) **ENTER**

 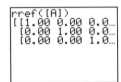

The following matrix with 1's along the diagonal and 0's in the upper and lower triangle should appear.

$$\begin{bmatrix} 1 & 0 & 0 & | & -.24 \\ 0 & 1 & 0 & | & .65 \\ 0 & 0 & 1 & | & -.13 \end{bmatrix}$$
which implies
$x = -.24$
$y = .65$
$z = -.13$

Activity: Solve the following system of equation:

$$X + 5y - 2z = 5$$
$$-4x + y = 3$$
$$-3x + 4y + z = -1$$

a. Use row-echelon form and back-substitution.
b. Use reduced row-echelon form.

E.3.3.: Use Inverse Matrix Method to Solve Simultaneous Equations

To solve a system of equations using inverses you will multiply the inverse of the coefficient matrix A by the constant matrix B. That is,

$[A]^{-1} [B]$ = [the solution matrix]

A solution will only exist if the inverse of matrix [A] exists. See finding the determinant of a matrix.

Example:
$2x + 3y - 4z = 2$
$-3x + 7y + 2z = 5$
$6x + 2y - z = 0$

Input the **coefficient matrix** as [A] a 3 x 3 matrix:

$$\begin{bmatrix} 2 & 3 & -4 \\ -3 & 7 & 2 \\ 6 & 2 & -1 \end{bmatrix}$$

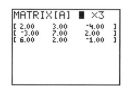

Input the **constant matrix** as [B] a 3 x 1 matrix:

$$\begin{bmatrix} 2 \\ 5 \\ 0 \end{bmatrix}$$

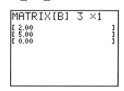

Find **inverse of [A]** which is $[A]^{-1}$
1. **MATRIX**
2. select **1:[A]**
3. **ENTER**
4. type x^{-1}
5. **ENTER**

Now multiply **[A]-1 [B]**

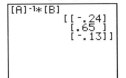

Hence, you will see the solution matrix which indicates x = -.24, y = .65 and z = -.13.

Activity: Solve the following system of equations using the inverse matrix method.

$$6x - 4y + 5z = 31$$
$$5x + 2y + 2z = 13$$
$$x + y + z = 2$$

E.3.4.: Addition/Subtraction Method
This is usually the first method taught in any Algebra course to solve simultaneous equations. To perform this method you will need to add/subtract/multiply rows.

Example: Find the solutions the following system of equations.
$$2x + 3y = 5$$
$$4x + y = 3$$

Input the augmented matrix in [C]

$$\begin{bmatrix} 2 & 3 & | & 5 \\ 4 & 1 & | & 3 \end{bmatrix}$$

First multiply rowA by -2 and add to rowB. User ***row+(-2,[C],1,2)**
1. MATRIX and MATH
2. select **:*row+(**
3. **ENTER**
4. type **-2,**
5. **MATRIX**
6. select **5:[G]**
7. **ENTER**
8. type **,1,2)**
9. **ENTER**

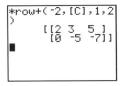

Input this matrix as [D].

Next multiply rowB by 1/5. Use ***row(1/5,[D],2)** which will give you the following matrix:

```
*row(1/5,[D],2)
    [[2 3  5  ]
     [0 -1 -1.4]]
```

Next input this matrix as [E]. Then multiply rowB by 3 and add rowB to rowA. Use ***row+(3, [E],2,1)**.

```
*row+(3,[E],2,1)
    [[2 0  .8 ]
     [0 -1 -1.4]]
```

Lastly, we just need to multiply rowA by 1/2 and rowB by -1 to get 1's in the diagonal and 0's in the upper and lower triangle. The resulting matrix thus is:

$$\begin{bmatrix} 1 & 0 & | & 0.4 \\ 0 & 1 & | & 1.4 \end{bmatrix}$$

The solutions are: x = .4 and y = 1.4.

Activity 1: Solve the following system of equations using the inverse matrix method:
   2x – 4y = 10,000
   3x – 7y = 40,000

Activity 2: Solve the following system of equations using any method:
   X + 2y – z + w = 2
   -2x – 3y – z – w = 4
   4x + 2y – 4z + 5w = 0
   -3x – 3y + 2z – 7w = 14

Activity 3: Encoding and Decoding a message

To encode a message, each letter in the alphabet is assigned an integer between 1 and 26. Ten each letter in a message is replaced by its integer counterpart with words separated by zeros.

Encoding:
The message
$$I\ AM\ LIZ$$

1. Is encoded as   9 0 1 13 0 12 9 26
2. And placed into a 2 x 4 matrix. The unscrambled message matrix **M** for the above code is

$$\begin{bmatrix} 9 & 1 & 0 & 9 \\ 0 & 13 & 12 & 26 \end{bmatrix}$$

3. M is scrambled by multiplying it by a 2 x 2 matrix **A** that has an inverse. Multiply **A x M**.

Let $A = \begin{bmatrix} 3 & 1 \\ 4 & 2 \end{bmatrix}$

4. Let **S = A x M** which is

$$\begin{bmatrix} 27 & 16 & 12 & 53 \\ 36 & 30 & 24 & 88 \end{bmatrix}$$

and send this scrambled message as   27   36   16   30   12   24   53   88

Decoding:
The receiver will get the scrambled message which does not seem to have any repeating letters. But we know that there are two I's in the original message. Hence, this message is more secure than an unscrambled message.

1. To decode this message find the inverse of A
2. Then multiply inverse of A by the matrix S.
3. You will have to send the recipient the key to the message, namely matrix A. Only then can the message be unscrambled.

$M = A^{-1} \times S = \begin{bmatrix} 9 & 1 & 0 & 9 \\ 0 & 13 & 12 & 26 \end{bmatrix}$

E.4.: More Matrix Operations

E.4.1.: Determinant
Det(will find the determinant of a square matrix.

$$A = \begin{vmatrix} 1 & 5 & 7 \\ 6 & 9 & 1 \\ 3 & 6 & 3 \end{vmatrix}$$

Find the determinant of matrix A, input matrix A then
1. **MATRIX**
2. **MATH**
3. select **1:det(**
4. **ENTER**
5. **MATRIX**
6. select **1:A** and type **)**
7. **ENTER**

 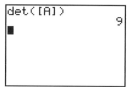

E.4.2.: Transpose
This operation will transpose any matrix. This means that you convert a matrix by changing rows to columns and columns to rows. Use **2:T**.

$$A = \begin{bmatrix} 1 & -2 \\ 6 & -3 \\ 3 & 1 \end{bmatrix}$$

Find the transpose of matrix A. First input matrix A, then
1. **MATRIX**
2. select **1:[A]**
3. **ENTER**
4. **MATRIX**
5. **MATH**
6. select **2:T**
7. **ENTER** and **ENTER**

 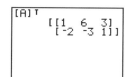

You will see that row 1 is now column 1, row 2 is column 2, etc.

E.4.3.: Dimensions of a Matrix

This operation will determine the dimension of your matrix. For example you would like to find the dimension of matrix A.

1. **MATRIX**
2. **MATH**
3. select **3:dim(**
4. **MATRIX**
5. select **1:[A]** and type **)**
6. **ENTER**

This tells you that your matrix A has 3 rows and 2 columns.

E.4.4.: Fills all elements with a constant

You can fill a matrix with let's say 3's. To do this use Fill(constant, matrix).
If you want to fill matrix C, a 2 x 3 matrix, with 3's.

1. **MATRIX**
2. **MATH**
3. select **4:Fill(**
4. **ENTER**
5. type **3,**
6. select **matrix C**
7. **ENTER**

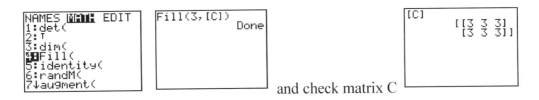

and check matrix C

All elements of matrix C have been filled with 3's.

E.4.5.: Identity Matrix

The function identity(dimension) returns the identity matrix which is square matrix. Create a 4 x 4 identity matrix.

1. **MATRIX**
2. **MATH**
3. select **5:identity(**
4. **ENTER**
5. type **4)**
6. **ENTER**

 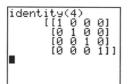

E.4.6.: Random Matrix: rand(rows, columns)

Creates a random matrix with the indicated rows and columns. The entries will be integers ≥ -9 and ≤ 9. Let's create a random matrix with 4 rows and 2 columns.

1. **MATRIX**
2. **MATH**
3. select **6:rand(**
4. **ENTER**
5. type **4,2)**
6. **ENTER**

 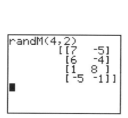

Practice Your Skills UNIT E

1. Find the solutions to the following system of equations:

   $$\begin{aligned} x + y - z - w &= 5 \\ 2x - 3y - 2z + w &= 7 \\ -7x + z - 3w &= -4 \\ 2y - z &= 1 \end{aligned}$$

   a. Find Matrix A, B, and X.
   b. Find the inverse of A.
   c. Find the transpose of B.
   d. Write the augmented matrix.
   e. Use ref method:
   f. Use rref method:
   g. Use the Inverse Matrix method:

2. Create a 4 x 4 identity matrix.

3. Multiply matrix A from above by the identity matrix.

4. Find the determinant of A.

5. Create a random 5 x 7 matrix.

6. Use the following matrices to perform the operations:
   a. Multiply D x C and C x D
   b. Negate matrix C
   c. Matrix D: Add 2 times row 2 to row 3.
   d. Matrix C: Multiply row 4 by -3.
   e. Matrix C: Multiply the matrix by -4.

   $$C = \begin{bmatrix} 1 & 2 \\ 3 & 2 \\ 1 & -1 \\ -2 & 0 \end{bmatrix} \quad D = \begin{bmatrix} 5 & 2 & -1 & 2 \\ -1 & 0 & 4 & 1 \\ 0 & 1 & 5 & 3 \\ 4 & 1 & -2 & -2 \end{bmatrix}$$

7. Encode and decode a message of your choice.

UNIT F : Other Topics

Equation Solver	**106**
Polar and Parametric Functions	**107-110**
Store Data and Pictures	**111-113**
Transmit Data and Programs	**114**
Create a Sequence	**115**
Practice Your Skills	**116**

F.1.: Equation Solver

We have learned various techniques to find x-intercepts of equations. There is another way to solve an equation, namely using the equation solver. It is applicable to equations of degree one and higher and finds all real roots. The method applies a sort of trial-and-error technique. Let's examine how this feature works.

F.1.1.: Find the real roots for the equation $x^2 - x - 6$.
1. **MATH**
2. **0:Solver**
3. **EQUATION SOLVER eqn:** $x^2 - x - 6$. (If you get another screen output just press the up arrow and it will return)
4. **ENTER**
5. Now try different values for x, i.e. $x = -1$.
6. Press **ALPHA … ENTER**.
7. It will tell you that $x = -2$, which is one solution.
8. Try another value for x, i.e. $x = 2$
9. Press **ALPHA … ENTER**.
10. It will tell you that $x = 3$, which is another solution.

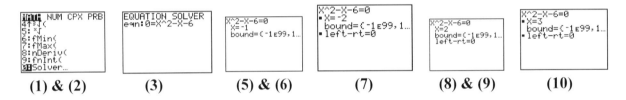

(1) & (2)   (3)   (5) & (6)   (7)   (8) & (9)   (10)

Hence, the solutions are $x = -2$ and $x = 3$. We can verify this fact by either graphing the equation in the Y= editor or use the factoring technique.

F.1.2.: Find the real solutions (roots or x-intercepts) for the equation: $x^3 - 3x - 1$
1. **MATH**
2. **0:Solver**
3. **EQUATION SOLVER eqn:** $x^3 - 3x - 1$ (If you get another screen output just press the up arrow and it will return)
4. **ENTER**
5. Now try different values for x, i.e. $x = -1$.
6. Press **ALPHA … ENTER**.
7. It will tell you that $x = -.347296…$, which is one solution.
8. Try another value for x, i.e. $x = 2$
9. Press **ALPHA … ENTER**.
10. It will tell you that $x = 1.879385…$, which is another solution.
11. What is the other real solution if there exists one? (Hint: It may help to graph the equation in the Y= editor mode first to see where the other zero approximately is located.)
    Answer: $x = -1.532088…$.

Practice: Find all real roots for the equation $x^4 - 5x^2 + 2$.

F.2.: Polar and Parametric Functions

F.2.1.: Polar Functions and Graphs

Polar equations are of the following format:
R = Asin(Bθ), where θ = radian measure of the angle.
Set your mode to Polar.
GRAPH will appear as follows.
WINDOW will be automatically changed to the following.
Use ZOOM Fit to view the graph

 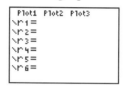  

Example 1: Graph **r1 = sin θ**. How many loops does the graph have?

Example 2: Graph **r2 = sin (3θ)**. How many loops does the graph have?

 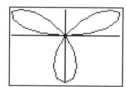

Example 3: Graph **r3 = sin (2θ)**. How many loops does the graph have?
Example 4: Graph **r4 = sin (4θ)**. How many loops does the graph have?
Example 5: Graph **r5 = cos (3θ)**.

 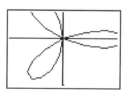

Example 6: Describe the differences between the graphs r = sin (3θ) and r = cos (3θ).

Activity 1: Predict how many loops r = sin (5θ) and r = sin (6θ) will have.
Activity 2: Predict how many loops r = cos (5θ) and r = cos (6θ) will have.
Activity 3: Describe the differences in the graphs when B is even and when B is odd for r = sin (Bθ).
Activity 4: What will happen if you graph r = 2 sin (2θ)?
Activity 5: Describe the differences in the graphs when A > 1 and A < 1 but greater than 0.
Activity 6: Graph r = cos (2θ) cos θ.
Activity 7: Graph r = cos (2θ) sin θ.

F.2.2.: Parametric Functions and Graphs

Parametric equations are of the following format:
   x = f(t), where t = is the independent variable and you can graph 3 components at the same time, namely x, y and t.
   Set your mode to Parametric.
   GRAPH will appear as follows.
   WINDOW will be automatically changed to the following.
You can input up to 6 different parametric equations.

 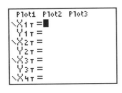 

Graph Parametric Functions:
Example 1: Graph **x = 2 and y = 3.**

Example 2: Graph **x = 2 and y = t.**

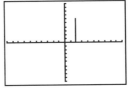

Example 3: Graph **x = t and y = 3.**

Example 4: Graph **x = 3t and y = 5t.** The linear equation is y = (5/3) x. Check!

Example 5: Graph **x = 3 + t and y = 4 − 2t.**

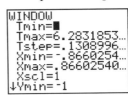 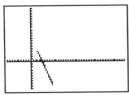

Complete the chart below:

T	x-coordinate	y-coordinate
0		
1		
2		
5		
20		

Find the linear equation for the above parametric equation:
Step 1: $X = 3 + t$
Step 2: Solve for t: $t = x - 3$
Step 3: $y = 4 - 2t$, where $t = x - 3$
Step 4: $y = 4 - 2(x - 3)$
   $Y = 4 - 2x + 6$
   $Y = 10 - 2x$. This is the equivalent linear equation for this parametric equation.

Example 6: The path of a ball. $X = 20t$ and $Y = -16t^2 + 30t$. x and y are measured in feet, t is measured in seconds.

  a. Graph the path of the ball.

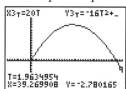

  b. Find the maximum height of the ball.
  c. After how many seconds does the ball reach maximum height?
  d. After how many seconds will the ball reach the ground?
  e. Find the quadratic equation for the above parametric equation.

Complete the chart below:

T	x-coordinate	y-coordinate
0		
.5		
1		
1.5		
2.0		

F.2.3.: Simulating Motion with Parametrics

Let's say you have two trains leaving the same destination heading for a city 1500 miles away. Train 1 travels at 250 mph, and train 2 travels at 400 mph, but leaves 2 hours later.
    a. Which train will reach the destination first?
    b. When will one train pass the other?

Note: Change your **MODE** to **Simul**.

 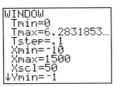

	Speed	Time	Distance
Train 1	250	t	1500
Train 2	400	t-2	1500

Parametric Equations:
X1T = 250 t
Y1T = 1
X2T = 400(t-2)
Y2T = 2
Graph and see what will happen.

Alternate Approach: Solve this problem graphically finding the intersection point.
Y = 250 x
Y = 400(x – 2)
Y = 1500
Answer the above questions:
    a. Which train will reach the destination first?
    b. When will one train pass the other?

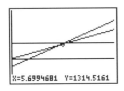 Zoom In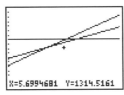

Activity: Simulate horse racing. Use 4 horses which are racing at the same time but at somewhat different speeds and starting at just slightly different times. Which will be the winner? You can let the students predict which horse will win.

F.3.: Store Data and Pictures

F.3.1.: Store Data

F.3.1.1.: Storing Variable Values

To store a value to a variable from the home screen or a program use the **STO→** key. Start at a blank line on your home screen.

Example: Store √30 as variable A.
1. type **√30**
2. press **STO→**
3. press **ALPHA** and then the letter of your variable, in this case **A**.
4. **ENTER**

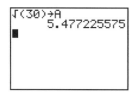

F.3.1.2.: Displaying a Variable Value
1. Enter the name of the variable on the home screen: In this case **ALPHA A.**
2. **ENTER** and you will see the value for A.

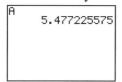

F.3.1.3.: Recalling
a. A Variable Value

Use Recall **RCL** to recall and copy variable contents to the current cursor location.
1. Press **RCL** (or 2nd STO→)
2. **ALPHA A**
3. **ENTER**

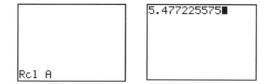

b. A Matrix

Let's assume you would like to recall matrix A.

1. Press **RCL** (or 2nd STO→)
2. **MATRIX ... select 1:[A]**
3. **ENTER**

 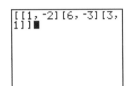

c. A Program

1. Press **RCL** (or 2nd STO→)
2. **PRGM ... EDIT**
3. Select the name of the program, i.e. **1:DYADVNCE**
4. **ENTER**

 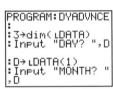   The entire program will appear!

F.3.2.: Store Pictures

Use **DRAW STO** to store a picture. You can store 10 pictures, Pic0, Pic1 through Pic9. Later you can superimpose the stored picture onto a displayed graph from the home screen or a program.

What is a picture? A picture on a calculator includes drawn elements, plotted functions, axes, and tick marks. A picture does not include axes labels, lower and upper bound indicators, prompts, or cursor coordinates. Parts of the display which are hidden by these items are stored with the picture.

F.3.2.1.: Create a picture first: Draw a balloon
1. Graph a circle with center (4,4) and radius 3.
   DRAW .... 9:Circle(4,4,3) ... ENTER
2. To draw a string to the balloon use
   DRAW ... A:Pen ...ENTER ... place your cursor at the starting point **... ENTER ...** draw your string the way you like it ...press **ENTER** when you are finished.
3. Turn your **CoordOff** and **AxesOff** and you will see something similar to this graph:

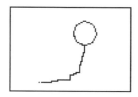

F.3.2.2.: Store the picture
1. **DRAW and STO**
2. select **1:StorePic**
3. pick a number for your picture, let's say **5**
4. **ENTER.** Your picture has now been stored.

F.3.2.3.: Recall your Picture:
1. **DRAW and STO**
2. select **2:RecallPic**
3. **ENTER ...** type number of picture **5.**
4. **ENTER**

F.4.: Transmit Data and Programs

To transmit data or a program from one calculator to another you will need the following:

Calculator 1: (Sender)
1. Two compatible calculators connected via cable
2. Go to **LINK**
3. **SEND** and select **PRgm**
4. **ENTER**
5. Select the program you wish to transmit, i.e. Dyadvnce, and press **ENTER**
6. Choose **TRANSMIT** and wait until you have calculator 2 set up.

Calculator 2: (Receiver)
1. Simultaneously set up the calculator you wish to transmit the data or program to.
    Go to **LINK**
2. Select **RECEIVE**
3. Press **ENTER** on receiver calculator 2.
4. Press **ENTER** on sender Calculator 1.

Calculator 1   **Calculator 2**

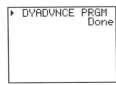 

Congratulations! You have now successfully transmitted program DYADVNCE from one calculator to the other.

F.5.: Create a Sequence

Create the following sequence: 1,3,5,7,9
Use **seq(X,X,starting point, end point, increments)**

1. **LIST** or 2nd STAT
2. **OPS**
3. Select **5:seq(**
4. **ENTER**
5. Input **X,X,1,9,2)**
6. **ENTER**

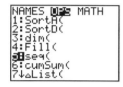 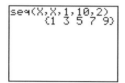

Activity: Create a sequence of multiples of 5 from 10 to 50

Activity: Create the following sequence: 100 to 2000, increments of 110 and store in L1.

Practice Your Skills UNIT F

1. Draw polar sunflowers.
    a. Graph $r = \sin(7/5)\theta$
    b. Graph $r = \sin(12/5)\theta$

2. Draw a 4-Leaf Clover: $r = 2\cos\theta \sin\theta$

3. Convert from rectangular to polar coordinate.
    Find r and θ. Use ANGLE and select 5:R>Pr(x,y) finds the radius and 6:R>Pθ(x,y) finds the angle in radians.

    a. Convert (-2,0) to polar coordinates: $r = 2$, $\theta = 3.1416$ radians $= \pi$
    b. Convert (-1,-3) to polar coordinates: $r = 3.16$, $\theta = -1.89$ radians
    c. Convert (3, 5) to polar coordinates: $r =$ ____ , $\theta =$ ____ radians.

4. Rock Climbing!
    You decided to go to Galyans with your students for a rock climbing adventure. Who will reach the top first? Create a parametric equation to model each students climbing speed and starting time. Graph to see who will win.

5. Create a picture of your choice and transmit to your partner's calculator.

6. Create a sequence of 111's up to 5000 and store in L1.

7. Use the Equation Solver to find all real roots of $y = x^5 - 6x^4 + 3x^3 - 5x^2 + 4x + 5$.

8. Draw a square using the equation $x^{20} + y^{20} = 10$.

9. Input a short program for computing the volume of a sphere, a pyramid, a circular cylinder and a triangular cylinder.

INDEX

Index

- A -

Absolute Value (abs), 14, 92
Access matrix, 88
Addition/Subtraction Method, 98
ALPHA, 18, 22, 106, 111
Angles, 15-17
    2nd Apps, 15, 16
    Catalog, 14, 18
    Alpha R, 18
    Ref, 18
AxesOn/AxesOff, 5

- B -

Back-Substitution, 95
Binomcdf(, 73, 75
Binompdf(, 73, 74
Box Plots, 55

- C -

CALC (2nd Trace), 39-44, 57
    Dy/dx, 39
    $\int f(x)dx$, 39
    Intersect, 39, 43, 44
    Maximum, 39, 42, 43
    Minimum, 39, 42, 43

    Par, 39
    Pol, 39
    Seq, 39
    Value, 39

Catalog, 14, 18
Circle(, 29, 33, 113
CLEAR, 52
ClrDraw, 29, 33
Coefficient Matrix, 97
Combinations, 67
Constant Matrix, 97
CoordOff/CoordOn, 5

Copy matrix, 87
Create a picture, 113
Create a sequence, 115
Cubic Regression, 60, 62
Curve Fitting, 60-65
    Cubic Reg, 60, 62
    Exponential Reg, 60, 64
    Linear Reg, 60, 61
    Logarithmic Reg, 60, 64
    Logistic Reg, 60
    Power Reg, 60
    Quadratic Reg, 60, 62
    Quartic Reg, 60, 62
    Sinusoidal Reg, 60

- D -

Dec, 21
Decoding a message, 100
Degree symbol/minute, 15, 16
Determinant, 101
Diagnostics, 19
    DiagnosticOn/ DiagnosticOff, 19, 61
Dimensions of a matrix, 102
Displaying variable values, 111
DISTR, 73 - 78, 81
Dot, 45
DRAW (2nd PRGM), 29-34, 81
    Circle(, 29
    ClrDraw, 29, 33
    DrawF, 29, 31
    DrawInv, 29, 32
    Draw Points, 29
    Draw Store (sto), 29
    Horizontal, 29, 30
    Line(, 29, 30
    Pen, 29
    Shade(, 29, 46

- D -

DRAW (cont.)
    Tangent(, 29, 31
    Text(, 29, 33, 34
    Vertical, 29, 30
Draw Sto, 113
DYADVNCE, 112

- E -

Edit, 50
Edit matrix, 86
Encoding a message, 100
Engineering Notation (eng), 23
Enter matrix from home screen, 87
Equation Solver, 106
Exponent, 20
    Fractional Exponent, 20
    3rd Root, 20
Exponential Regression, 60, 64
ExprOn/ExprOff, 5

- F -

Factorial, 65, 68
Fcdf(, 73
Fpdf(, 73
Fill a matrix, 102
Float, 2
Format Settings, 4-5
    AxesOn/AxesOff, 5
    CoordOff/CoordOn, 5
    ExprOn/ExprOff, 5
    GridOff/GridOn, 4
    LabelOn/LabelOff, 5
    PolarGC, 4
    RectGC, 4

Fractions (frac) and Decimals (dec), 21
Full Screen, 3
Func, 6-7, 10

- G -

Geometcdf(, 73, 79
Geometpdf(, 73, 78
GRAPH, 28, 53, 55
Graphing Modes
    Func, 6-7, 10
    Par, 6-7, 10
    Pol, 6-7, 10
    Seq, 6-7, 10
GridOff/GridOn, 4
G-T split, 3, 10

- H -

Histogram, 54
Horizontal, 29, 30
Horiz Screen, 3

- I -

Intersect, 39, 43, 44
Identity matrix, 103
Inverse of a matrix, 93, 97
InvNorm(, 73

- L -

LabelOn/LabelOff, 5
Line(, 29
Linear equation, 109
Linear Regression, 19, 60, 61
Link, 114
Lists, 50-52, 115
    ClrList, 50
    Standard Deviation, 51
    Sort, 52

- L -

Logarithmic Regression, 60, 64
Logic, 24
Logistic Reg, 60

- M -

MATH Menu, 14, 20, 21, 51, 65, 67-72, 89, 95, 106
Matrix, 86-90
    Access matrix, 88
    Coefficient Matrix, 97
    Constant Matrix, 97
    Copy matrix, 87
    Edit matrix, 86
    Enter matrix from home screen, 87
    Row addition, 89, 90
    Row multiplication, 90
    Row Swap, 89
    Store/Recall a value, 88
    View matrix, 87
Matrix Operations, 91-94, 101-103
    Absolute value of a matrix, 92
    Add and subtract matrices, 91
    Determinant, 101
    Dimensions of a matrix, 102
    Fill a matrix, 102
    Identity matrix, 103
    Inverse of a matrix, 93, 97
    Multiply matrices, 91, 92
    Negate a matrix, 92
    Powers of a matrix, 94
    Random matrix, 103
    Relational operations, 94
    Round matrix values, 93
    Transpose, 101
Maximum, 39, 42, 43

Memory (mem) Storage (sto) and Retrieval, 22
    Alpha A, 22
    Check available memory – Mem Mgmt/Del, 22
Minimum, 39, 42, 43
Mode Settings, 2-3, 6, 15
    Degree, 2, 16
    EE, 23
    Eng, 23
    Float, 2
    Full Screen, 3
    Graph, 3
    G-T split, 3, 10
    Horiz Screen, 3
    Polar, 107
    Radian, 2, 15-17,
    Sci, 23
    Simulation, 110
    Split Screen, 3
    2nd Mode, 15, 16, 23, 24
Modified Box Plots, 55
Multiply matrices, 91, 92

- N -

nCr, 65
nPr, 65
Negate a matrix, 92
Normalcdf(, 73, 80
Normalpdf(, 73, 80
Normal Probability Plots, 56
Num, 14

- O -

Order of Operation, 17
OPS, 115

- P -

Par, 39
Parametric Functions, 108 - 110
    Graph parametric functions, 108,109
    Simulating motion, 110
Pt-On(, 29
Pt-Off(, 29
Pt-Change(, 29
Pxl-On(, 29
Pxl-Off(, 29
Pxl-Change(, 29
Pxl-Test(, 29
Pen, 29
Permutations, 66
Poissoncdf(, 73, 77
Poissonpdf(, 73, 76
Pol, 39
Polar Functions, 107
    Sin, asin, cos, 107
PolarGC, 4
Power Reg, 60
Powers of a matrix, 94
PRB, 65-72
PRGM, 112
Probability Computations (prb), 65-68
    Factorial, 65, 68
    Number of combinations, 65, 67
    Number of permutations, 65, 66
    Random-number generator, 65
    Random # from Binomial Dist, 65
    Random # from Normal Dist, 65

Probability Distributions (2nd Vars), 73-81
    Distr menu:
        Binomcdf(, 73, 75
        Binompdf(, 73, 74
        Fcdf(, 73
        Fpdf(, 73
        Geometcdf(, 73, 79
        Geometpdf(, 73, 78
        InvNorm(, 73
        Normalcdf(, 73, 80
        Normalpdf(, 73, 80
        Poissoncdf(, 73, 77
        Poissonpdf(, 73, 76
        Tcdf(, 73
        Tpdf(, 73
        X^2cdf(, 73
        X^2pdf(, 73
    Draw menu
        ShadeNorm(, 73, 81
        Shade_t(, 73
        ShadeX^2(, 73
        ShadeF(, 73

- Q -

Quadratic Regression, 60, 62
Quartic Regression, 60, 62

- R -

Radians, 15-17
RAM, 22
Rand, 65, 69, 70
RandBin, 65, 69, 72
RandInt(, 65, 69, 70
RandNorm(, 65, 69, 71
Random matrix, 103

- R -

Random integers, 69, 70
Random numbers, 69
Random # from Binomial dist, 69, 72
Random # from Normal Dist, 69, 71
Random-number generator, 65
Random # from Binomial Dist, 65
Random # from Normal Dist, 65
RCL, 111, 112
RecallGDB, 29
RecallPic, 29
Recalling a matrix, 112
Recalling a program, 112
Recalling variable values, 111
Receive, 114
RectGC, 4
Reduced Row-Echelon Form, 96
Relational operations, 94
Round matrix values, 93
Row addition, 89, 90
Row-Echelon Form, 95
Row multiplication, 90
Row Swap, 89

- S -

Scalar Multiplication, 91
Scientific Notation (sci), 23
Set Factors, 38
Sequence, 39, 115
    Create a sequence, 115
Shade, 45-46
    Graph styles – line, thick, above, below, path, animate, dotted, 45
ShadeNorm(, 73, 81
Shade_t(, 73
ShadeX^2(, 73
ShadeF(, 73
Simulation, 69, 72, 110

Sin/Asin/Cos, 107
Sinusoidal Reg, 60
Solving Simultaneous Equations, 95 -100
    Addition/Subtraction Method, 98
    Back-Substitution, 95
    Encoding and Decoding messages, 100
    Inverse Matrix Method, 97, 98
    Reduced Row-Echelon Form, 96
    Row-Echelon Form, 95
Sort, 52
Split Screen, 3
Standard Deviation, 51
Stat Analysis, 57-59
    Var Stats, 57
    2-Var Stats, 59
    Weighted Mean, 58
Stat Plots, 28, 52-56
    Box-plots, 52, 55
    Histograms, 52, 54
    Modified Box Plots, 52, 55
    Normal Probability Plots, 52, 56
    Scatter Plots, 52, 53
    XY Line charts, 52, 53
    Y-Editor, 52
StoreGDB, 29
StorePic, 29, 113
Store Data (STO), 87, 111-112
    Storing variable values, 111
    Displaying variable values, 111
    Recalling variable values, 111
    Recalling a matrix, 112
    Recalling a program, 112
Store/Recall a value, 88
Store Pictures, 113
    Create a picture, 113
    Store picture, 113
    Recall a picture, 113

- T -

Table Settings, 8-10
    TblSet, 8-9
    2nd Graph, 8-9
    2nd Window, 8-9
Tangent, 29, 31
Tcdf(, 73
Test, 24
Text(, 29
Tpdf(, 73
Transmit Data and Programs, 114
    Link – send - transmit, 114
    Link - receive, 114
Transpose, 101
Trig, 31

- V -

Value, 39, 40
VARS, 57
Var stats, 57-59
View matrix, 87

- W -

Weighted Mean, 58
Window Settings, 6-7, 28
    Func, 6-7
    Par, 6-7
    Pol, 6-7
    Seq, 6-7

- X -

X^2cdf(, 73
X^2pdf(, 73
XY Line charts, 52, 53

- Y -

Y-editor, 52

- Z -

Zero, 39, 41
ZOOM, 35-39
    Zbox, 35, 36
    Zoom In/ Out, 35, 37, 110
    Zdecimal, 35, 37
    Zinteger, 35
    Zprevious, 35
    Zsquare, 33, 35, 38
    Zstandard, 35, 36, 38, 40, 53
    Ztrig, 35, 38, 40
    ZoomStat, 35, 53, 55
    ZoomFit, 35
    Zoom Memory, 35, 37
    Set Factors, 38

NOTES:

NOTES:

Adventures In Education, Inc.
3460 S Fletcher Ave Ste 205
Fernandina Beach, FL 32034

Adventures In Education also offers graduate
level distance learning courses in Math and Science.
To review our offerings please visit our website at
http://www.adventures-in-education.com

To order copies of this book contact us at
Email: adventuresinedu@earthlink.net
Tel. 630-877-4006